ROUTLEDGE LIBRARY EDITIONS:
POLLUTION, CLIMATE AND CHANGE

Volume 15

OIL POLLUTION CONTROL

OIL POLLUTION CONTROL

SONIA M. ZAIDE

LONDON AND NEW YORK

First published in 1987 by Croom Helm Ltd

This edition first published in 2020
by Routledge
2 Park Square, Milton Park, Abingdon, Oxon OX14 4RN

and by Routledge
52 Vanderbilt Avenue, New York, NY 10017

Routledge is an imprint of the Taylor & Francis Group, an informa business

British Library Cataloguing in Publication Data
A catalogue record for this book is available from the British Library

ISBN: 978-0-367-34494-8 (Set)
ISBN: 978-0-429-34741-2 (Set) (ebk)
ISBN: 978-0-367-36296-6 (Volume 15) (hbk)
ISBN: 978-0-367-36432-8 (Volume 15) (pbk)
ISBN: 978-0-429-34593-7 (Volume 15) (ebk)

Publisher's Note
The publisher has gone to great lengths to ensure the quality of this reprint but points out that some imperfections in the original copies may be apparent.

Disclaimer
The publisher has made every effort to trace copyright holders and would welcome correspondence from those they have been unable to trace.

OIL POLLUTION CONTROL

SONIA ZAIDE PRITCHARD

CROOM HELM
London • Sydney • Wolfeboro, New Hampshire

Croom Helm Ltd, Provident House, Burrell Row,
Beckenham, Kent, BR3 1AT

Croom Helm Australia, 44-50 Waterloo Road,
North Ryde, 2113, New South Wales

British Library Cataloguing in Publication Data

Pritchard, Sonia Zaide
Oil pollution control.
1. Oil pollution of the sea — Prevention
— History
I. Title
628.1'6833'09162 GC1085

ISBN 0-7099-2094-6

Croom Helm, 27 South Main Street,
Wolfeboro, New Hampshire 03894-2069, USA

Library of Congress Cataloging in Publication Data
applied for:

Printed and bound in Great Britain by
Biddles Ltd, Guildford and King's Lynn

CONTENTS

PREFACE

Possibly no other environmental problem has received as much international action, public attention and scholarly study as the control of oil pollution at sea. The record of these efforts dates back to the early 1920s, when oil replaced coal as the main source of fuel for technological progress. From primary documents and other sources, including previously confidential papers of the British and American officials involved and interviews and correspondence with contemporary policymakers in government, industry and pressure groups, it is now possible to present a comprehensive survey of the story of oil pollution control.

I have tried to analyse how policy has been made based on primary records (where these are available) and on my own experience as a participant in an international conference on marine pollution. After having studied the documents on early attempts to control oil pollution, I was impressed with the way in which the issues and alignment of interests seem to recur whenever the problem is considered. The historical principles certainly repeated themselves, but it has been left for individuals and groups to perceive and act upon them.

The oil pollution story starts with the draft Washington convention produced by the world's first environmental conference in 1926, and it ends with the 1973 Marine Pollution Convention, the world's first umbrella treaty to deal with pollution by ships at sea. This survey of oil pollution control policy aims, firstly, to correct various fallacies connected with government or industry action, and secondly, to spur interest in solving the remaining problems which prevent the satisfactory control of oil pollution.

An analytical framework for the study of environmental problems is also provided in the concluding chapter.

This book was partly based on my Ph.D. thesis at the London School of Economics and Political Science in 1975. So many people and organisations have helped in the research and preparation of this book that it is not possible to acknowledge all of them. I wish to thank the following especially: my parents, Greg and Lily Zaide, who financed my studies and gave loving support through the difficult periods; Alan M. James, my LSE supervisor and now Professor at Keele University; David M. Croom of Croom Helm; and the British Council in Manila.

Like the thesis, I dedicate this book to my children, Robbie and Michelle, with Christ's love that is the perfect bond of unity.

1

EARLY LOCAL AND NATIONAL ACTION

When oil replaced coal as the more efficient and convenient fuel at the start of the 20th century, oil pollution in ports and coastal waters inevitably followed the trade. Local and national officials made various attempts to curb the problem of oily wastes from ships. But in the first place they had to be convinced of the need to take action, and, secondly, the co-operation of the oil and shipping industry had to be deferentially solicited before the principle could be established that technological progress should be accompanied by waste management policies.

During and immediately after the First World War, widespread oil pollution from ships – mostly due to highly flammable gasoline discharges – caused fire hazards and damage in ports and health risks in coastal resorts. Even as the Allies floated to victory on a wave of oil, the British Admiralty and Ministry of Shipping felt compelled to issue wartime instructions to control oily discharges from ships. These wartime regulations, though not having the force of a Defence of the Realm Act, were issued in 1918 instructing shipmasters to discharge oily ballast and cleaning water from their ships outside of the three-mile limit of territorial waters and to take precautions against leakages of oil when loading, unloading, or re-fueling at port.[1]

After the war, the power to curb oil pollution reverted to port officials rather than national authorities. Without national supervision, the situation deteriorated, inviting mounting complaints from officials in ports affected by harbour fires and from fishermen and resort owners experiencing damage to fisheries and amenities. The aggrieved communities initially endeavoured to deter pollution in ways which would not frighten away the very source of their livelihood. These delicate measures ranged from verbal appeals to shipowners and shipmasters, distribution of leaflets, posting of billboards, and only finally by direct prosecution of masters who were caught flagrantly discharging oily wastes within harbour areas.

But even if they had local by-laws to protect their areas, port officials noted that their powers to curb pollution in harbours dated to an era when oil ships were the oddity, not the rule, and the penalties were so ridiculously low as to invite defiance. In Britain, the by-laws, for harbour pollution dated to King Henry VIII's reign in 1543 and carried a maximum fine of £5[2]. Outside of ports, coastal town councils were not empowered to prosecute ships for oil pollution, so that in many ways the public did not have the barest degree of protection.

In the United States, then the centre of the oil trade, a spate of harbour fires caused by oil focused attention on the vulnerability of New York harbour. The entire harbour was said to be potentially destructible if a flare or an electric wire had accidentally been dropped in the river. With over $11.5 billion in annual trade at stake, fire underwriters and New York harbour officials secured in 1921 the designation of a 25-mile zone within which it was prohibited to discharge oily wastes from ships.[3]

Britain Pioneers National Legislation on Oil Pollution

The more fertile field for national, as opposed to local, legislation on oil pollution lay in British. Whilst a number of port regulations in the world had preceded Britain's national law on oil pollution, the former were isolated rules and did not have uniformity or the full weight of national authority behind them.[4]

Following their then acknowledged lead in maritime legislation, officials of the British Board of Trade's Mercantile Marine Department initiated discussion on a bill for Parliament. In January 1921, Charles Hipwood, head of the Marine Department, presided over a series of informal conferences between oil and shipping industry representatives and dock and harbour officials. Port officials went to the talks with the support of two notable British environmental groups, namely, the Royal Society for the Protection of Birds and the Royal Society for the Prevention of Cruelty to Animals. But the Board considered it 'unnecessary and perhaps divisive' to invite or include members of these animal protection groups and coastal town councils who were agitating for the strongest possible measures against oil pollution.[5]

The discussions brought out their differences on two matters – the necessity for a new law on oil pollution and the scope of such proposed legislation. At the earliest stage in the negotiations, it became evident that only government mediation could reconcile the differences of opinion between the commercial interests and the port authorities, since each had come to an entirely opposite view on the problem. Significantly, the discussions held at the Board of Trade on January 19-21, 1921, were only the first indications of a continual dichotomy in environmental legislation negotiations.

On the principle of a new legislation, port interests maintained that only a bill by Parliament could effectively prevent oil pollution by ships. Their representatives pointed out that the present by-laws were very unsatisfactory, and in any case the maximum penalty which could be levied was quite inadequate. On the other hand, the oil and shipping interests resisted the proposal of stiffer penalties and general legislation since they considered it unfair to saddle British shipping with new regulations that may hamper the domestic oil and shipping trade. From the confidential records of the discussions it appears that Hipwood's personal initiatives and determination paved the way toward new legislation on oil pollution. Maritime interests were reminded that their own ships were liable to suffer in case of the petrol wastes in harbours catching fire, and with the advent of more tankers and oil-fueled ships, the situation in ports was bound to become more serious.

The parties then proceeded to discuss several means of preventing oil pollution, ranging from higher penalties alone to technical innovation as well. All parties agreed that the best way of preventing the oil pollution problem was the provision of port reception facilities – receptacle barges or land storage tanks – into which the oily water from ships or tankers could be transferred. In such a way, the problem of ashes from coal had been dealt with, when coal pollution had been the major coastal trouble.

However, the provision and use of reception facilities for oily wastes were yet a novelty and proved exceedingly controversial. Port officials were reluctant to invest in separator barges or land storage tanks, especially since they regarded the oil problem as a matter for shipowners and oil companies to deal with. For their part, shipowners and oil companies were unwilling to include port reception of oily wastes as part of their normal operations due to the loss of turn-around time or the imposition of additional port and customs levies. Hence, any absolute prohibition against routine discharges of oily wastes from ships involved a substantial modernisation of ships and ports which seemed unacceptable to a highly competitive industry.

In an attempt to break the deadlock, British interests settled for the zone system as a compromise. They agreed to revive the practice urged upon them by the Admiralty during the First World War – to pump oily wastes only outside the three-mile limit from shore. This zone system, regarded during the war as a precaution against port fires, was to be imposed on shipping, but this time under statutory penalty.

However, oil and shipping interests maintained that the best method of preventing oil pollution was to deposit oily wastes in port. They were prepared to consider port reception of oily wastes as part of their normal operations and to pay for the use – but not the provision – of such facilities.

Dock authorities, who could have provided such reception facilities, unfortunately refused to be under any legal obligation whatsoever to do so. Instead, they issued a collateral undertaking to 'examine' the question of providing such facilities in large ports, an assurance deemed acceptable to the maritime interests at the time.

Thus, after having settled the question of port reception facilities, the main technical bone of contention in any oil pollution discussion, the parties agreed to remain within the four corners of the new law than for them to consider new vessel design and construction.

The Oil in Navigable Waters Act of 1922, which took effect from 1st January 1923, became the the first national law against oil pollution. It brought temporary relief of pollution in ports and harbours by prohibiting, under maximum fine of £100, a ship or land installation from causing oil or oily wastes to escape into British territorial waters.

But the full rigour of the law was kept at a distance. The main responsibility for prosecution was left with dock officials in practice. Whilst the Board of Trade and other ministries were empowered to delegate the authority to prosecute, a secret Board memorandum noted that they had no intention of using such powers if they could help it.[7] Furthermore, Board of officials concluded that the mere imposition of penalties would not check oil pollution, for cases would be exceedingly difficult to detect and to prove.

The best means of curbing oil pollution – the provision and use of port reception facilities – was founded on a gentlemen's agreement and was not made part of the new Act. Ironically, this arrangement led to a false sense of security. Little use was made or port facilities, even where they had been provided at minimal costs, for shipmasters preferred to avail of the zone system.

Indirectly, the 1922 Act had the effect of diverting oil pollution away from ports onto coastal areas in the vicinity of ports. Evidently, oily discharge from ships outside of the three-mile limit yet drifted to coastal areas even where no ports existed, e.g. the Isle of Wight. This novel problem was manifested in the complaints made after the 1922 Act. Complaints from dock officials, previously the more numerous complainants, virtually vanished whilst grievances from coastal town councils considerably mounted.

It must also be noted that the 1922 British Act carried little penalty for violators. By 1931, it was found that only 43 out of 54 cases had successfully been prosecuted; a pittance of £1 was recorded as a fine for one offence. The Board had information that all oil slicks drifted to shore, and that in 75% of the time these spills were made within the three-mile limit.[8] Rather than act upon this information, however, Board officials aspired for similar legislation in other countries. When the problem of oil pollution was elevated at Parliament, Sidney Webb, the Board of Trade President, could only reply that Britain was committed to international action.[9] Ironically, the first national anti-pollution legislation virtually promoted the certainty of pollution, albeit at some distance from shore.

The United States Stirs Up International Interest

American officials also realised that they had relatively little protection against oil pollution. The United States was then the chief source of oil, a position presently taken by the Middle Eastern countries. To world markets at the time, the United States supplied 60% of all oil and oil products. American attitude towards any oil policy was crucial.

Strong lobbies of private and commercial groups, notably fire underwriters and the National Coast Anti-Pollution League, urged oil pollution legislation in the United States. From 1922 to 1924, six major bills were presented to Congress after the passage of a joint Congressional resolution urging the President to summon an international conference on oil pollution.[10]

On 7th June 1924, at the height of the summer season when pollution complaints usually increased, Congress passed the Oil Pollution Act of 1924.[11] As compared to the 1922 British Act, the 1924 U.S. Act had more onerous provisions and was more vigorously enforced; as such it became more effective than the British legislation. In the first place, the

U.S. Coast Guard, apart from port officials, were given legal enforcement powers. Secondly, the definition of oil was broader than in the British Act. And finally, aside from a maximum fine of $2,000, which was a stiffer deterrent than in the British Act, the 1924 U.S. Act also made oil pollution a penal offence.

In other ways, the 1924 U.S. Act also adopted the zone system of banishing oil pollution to outside territorial waters, and only its stricter enforcement contributed to some improvement on the problem in American waters. The situation was not yet deemed entirely satisfactory to both British and American policymakers, and they encouraged similar national legislation elsewhere and also the possibility of an international agreement.

Following the 1922 joint Congressional resolution, President Warren G. Harding authorised U.S. Secretary of State Charles E. Hughes to begin the preparations for an international conference on oil pollution. Secretary Hughes appointed an ad hoc Interdepartmental Committee on Oil Pollution, with the State Department's economic adviser, Dr. Arthur N. Young, as chairman. The Committee was tasked with the collection of information on the effects and the control of oil pollution, and it was authorised to call an international conference only after it had determined ways to control oil pollution that were both practical and economically viable.[12]

Anglo-American Co-Operation on Oil Pollution Control

British and American co-operation on oil pollution control in the 1920s and 1930s, though proof of close policy co-ordination, eventually diverged in terms of interest and leadership. In arranging for the world's first international conference on oil pollution, the Americans believed that sufficient technical and practical information would commend oil pollution control to other countries. The British, however, were not entirely convinced that maritime interests would agree to more effective preventive measures involving radical alteration of ships and ports. In time, the

Americans abandoned the product of the joint Anglo-American labors, whilst the British sustained and even initiated efforts for international action.

Soon after the establishment of the Interdepartment Committee in 1922, American officials requested British co-operation in gathering information on oil pollution. With great enthusiasm, British officials clutched at this opportunity for co-operation due to two reasons. Firstly, they wanted to ensure that the Americans would not pull another surprise on them similar to the stiffer fines and penal provisions of the 1924 Oil Pollution Act, which were imposed on British shipmasters caught discharging waters within the U.S. zone.[13] Secondly, securing international action through the proposed Washington conference was preferred to taking stronger measures on the domestic front, and British officials wished to avoid pressure from environmental groups at home. Again, Hipwood took a keen personal interest in the matter and wanted the two countries to agree on the broad lines of a common policy.[14]

There were several opportunities for the Americans and the British to co-ordinate their policies. In March 1923 two naval attachés in London were detailed to study the Board's records and to meet other officials in order to study the legal and administrative measures and the mechanical devices to prevent oil from ships.[15] In July 1923, Joseph B. Frelinghuysen, a former Senator for New Jersey, came to London as the personal emissary of President Harding to attend informal Board discussions and meetings with government and shipping officials. Frelinghuysen had just retired as Senate President, a position which he held in 1909-10 and then again from 1917-23, and actively campaigned for oil pollution control among his numerous crusades. He founded the U.S. National Coast Anti-Pollution League, and in 1926 presided over the Washington Conference on oil pollution. Hipwood attached great importance to Frelinghuysen's visit, since he was one of the most prominent Americans in the anti-pollution campaign.[16]

On technical matters, British assistance had unequalled value to the American investigations on devices to prevent oily wastes from ships. From 1921 to 1924 the Board made independent investigations into the techniques and devices for separating oily wastes in ships, and they had come to the conclusion that several experimental or working separators would enable ships to retain and to separate oily wastes from cleaning or ballast water in their holds.[17] When the Board was informed at one stage that the American investigations had reached an impasse due to failure to discover an economic device to enable ships to control oily wastes, extensive information, even including trade secrets, was forwarded, to them by the Board and the British Chamber of Shipping.[18]

Indirectly, the American investigations helped British officials in deflecting attention from the inadequacy of the latter's domestic policy. In the summer of 1924, the Board again faced public agitation for further controls against oil pollution, and they diplomatically nudged the Americans about holding the international conference.[19] After being informed that the reason for the delay was a technical impasse, the Board sent all information they had about the shipboard separation of oily wastes. In 1925, domestic pressure again caused the Board to ask the Americans to expedite the conference, a request which the U.S. government denied due to the ongoing investigation.[20]

The American Inter-Departmental Committee Report

The American Inter-departmental Committee on Oil Pollution of Navigable Waters drew representatives from the Departments of State, Commerce, War, Treasury, Interior, and Agriculture, as well as the U.S. Shipping Board; and the American Petroleum Institute for the oil companies, and the American Steamship Owners Association for the shipowners. Their terms of reference, coming from Secretary of State Hughes, emphasised that they must summon an international conference only when they had practical suggestions to prevent the oily problem.[21] It was not until four years later that a satisfactory report could be rendered and the date for the conference fixed.

In September 1922, the Bureau of Mines, which had originally invented the "cracking process" to produce gasoline from crude oil, offered its expertise to investigate the technical features of oil pollution. In accepting the offer, Secretary Hughes stressed that the investigation must not only be comprehensive but also result in discovering the 'economic practicability of the methods and devices' to be employed.[22] The Bureau was urged to render a report within three months. But the task proved more complex than had been anticipated so that the Bureau was forced to admit, after several 'partial reports', that they could not immediately recommend any practical anti-pollution method which was at the same time a viable economic proposition.[23]

For ships using oil fuel, the Bureau recommended the re-cycling of fuel oil wastes onboard the ship itself. Existing separators in the market, especially those used by British passenger liners since 1924, were said to be easily capable of separating and re-cycling fuel oil from ballast water.[24] For oil tankers, however, the Bureau noted that the heavy sludge and bulky residues in cargo tanks could not easily be separated. The effective rate of separation of most devices, in tons per hour, could not handle the volume of tanker oily mixtures. In order to clean the tanker completely, the procedure was best done at the port of discharge or after delivery of the oil cargo and before the tanker sailed with ballast water in tanks. Indeed, this was the practice for Dutch tankers bringing oil from the East Indies to Amsterdam[25] However feasible the procedure was, it did not seem economical. Tankers would have to lie idle during cleaning in the port, and oil terminals would have to undergo extensive modification, so that American oil companies steered research away from this direction.[26].

It became imperative to choose another means of prevention at the source – retention and separation of oily wastes within the tanker itself, and to this end various bureaux busied themselves. The investigations achieved a technical breakthrough with the co-operation of the Standard Oil Company of New Jersey. On 25 July 1925, a company tanker, the *Charles Pratt,* sailed from New York with 120,000 barrels of Regan Country crude oil and arrived in Texas City, Texas, with 350 barrels recovered during the voyage by means of a skimming device which had especially been fitted for the experiment.[27] A scientist from the Bureau of Standards, Dr. David V. Stroop, was present throughout and gave a full report to the Committee.[28]

However, if the United States desired to recommend the complete retention and separation of oily wastes onboard ships and tankers, it was necessary to alter legislation in related matters—customs dues on the retained oil onboard ships, tonnage measurement excluding the separator apparatus, and vessel construction, standards for new or old ships. Moreover, the new series of legislation to prevent oil pollution by ships would have to be acceptable to all major maritime nations.

The Report of the Committee, and the diplomatic invitation to attend the 1926 Washington Conference on oil pollution, trumpeted the advantages of shipboard retention and separation of oily wastes.[29] American shipping companies acknowledged that the recommended measures were revolutionary, but they gave their support.[30]

Public Opinion and Environmental Policy

In the 1920s and 1930s, oil pollution hardly attracted the same degree of controversy as it does today. But the issue was present, so that government officials or industrial interest could ill-afford to neglect the claims of a band of dedicated advocates, who, possessing a legitimate grievance, endorsed the problem for official action as a moral crusade. Public officials and industrialists were persistently reminded by environmentalists of their social obligation to protect resources, coastal amenities, birds, fish, and the balance of nature. 'Gallon for gallon,' it was said that oil was the most devastating threat to marine ecology, coastal amenities, and coastal resources.

The groups used several methods, including barrages of letters to or personal calls on officials and shipowners, hortatory resolutions passed at assemblies, parliamentary questions, and publicity. Their two-pronged attack aimed to influence both governments and maritime interests to realise and to act against the 'evils' of oil pollution. Since they did not have the substantial attention or even sympathy which presently can be mustered for such causes, environmentalists waged an uphill battle against powerful commercial interests and official inertia. They played the role of the aggrieved party to the hilt. Of course, official policy turned the 'wrong' way a countless number of times, but environmentalists made an impact in three ways. Firstly, they provided a human dimension to a problem which seemed a dreary and unrewarding task to officials and industrialists. Secondly, by supporting positions in advance of what was acceptable to governments or industry, they made it possible for policymakers to consider radical proposals and to return to such measures when compromise or palliatives had failed. Thirdly, pressure groups created a climate of opinion favourable to change.

In the United States, the most powerful oil pollution lobby in the 1920s was organised by the National Coast Anti-Pollution League, which was founded in August 1922 at Atlantic City, New Jersey, to urge congressional passage of anti-pollution legislation. But after having achieved passage of the 1924 Act and obtaining good results from it, the League was disbanded. This situation confirmed official impression, at least in the U.S., that public complaints had 'died down' after passage of new legislation.[31]

British groups were not as ephemeral. Port authorities had lain low after passage of the 1922 Act, it is true, but coastal town councils, sea fishery boards, women's federations, the Royal Society for the Prevention of Cruelty to Animals (RSPCA), and the Royal Society for the Protection of Birds (RSPB) waged a sustained campaign for greater protection against oil pollution.[32]

British environmentalists established a model campaign to badger politicians and bureaucrats, who found themselves in a state of perpetual unease instead of smug complacency. When letters from town councils about a particular complaint failed to secure results, it was almost certain that this would be followed by questions in parliament, news reports, or, even at one stage, a petition signed by 18,000 residents of the Isle of Wight to the Prime Minister.[33]

Scientific evidence on the actual extent and damage of oil pollution was not always conclusive, and from a policy-making point of view such findings produced a degree of uncertainty, not to mention a convenient excuse for inaction. Environmentalists however contested these data. When informed at one stage that scientific studies on oil at sea did not always support the case for stronger measures, the RSPB offered to put up a museum exhibit of oiled birds for conference delegates.[34]

Even more significant differences prevailed between those who wished oil pollution to be controlled at all costs and those whose livelihood would be disrupted by strict controls. The relations between these two groups fluctuated between co-operation and hostility. When the RSPB declared oil pollution to be the greatest menace to birds, next to that of the trade in bird plummage and hunters, shipowners invited their representatives to observe at the International Shipping Conference meeting in 1924. Lord Montagu of Beaulieu and Hugh Gladstone represented the bird protection societies, but they had no vote and eventually disassociated themselves from the findings of the shipowners.[35] In 1931, in the wake of oil pollution covering a distance of eight miles off the Welsh coast, the RSPB brought the owners of the tanker *Ben Robinson* to court, and the company was fined £25 in celebrated proceedings published in the *Pembrokeshire Telegraph* of 22 January 1931.

Officials, industrialists and environmentalists differed on the question of the measures to effectively prevent the oil pollution problem. Industrialists equated good results from the observance by their ships of good housekeeping measures and national laws, despite their resistance to such laws until passage was beyond forestalling. Environmentalists felt that conditions were far from satisfactory, and that both maritime interests and governments should advance toward total prohibition measures.

In 1935, the Attorney-General, Sir Thomas Inskip, wanted measures for accidental pollution, after he personally saw the damage from oil caused by the wreck of the steamship *Letitia* off south Devon, and with his holiday guests had to humanely exterminate sixty oiled seabirds.[36] Both the Admiralty and the Board took the position however that, salvage difficulties apart, there was 'time enough to worry about the wrecks when we have a satisfactory control over the ships afloat.'[37]

It was also suggested by environmentalists that British officials consider repealing the 1922 Act to provide for the compulsory provision of either port reception facilities or shipboard separators, or the extension of the width of territorial waters beyond the three-mile limit.

Trade and other delicate ingredients of national interest were mixed in the melting pot of policy. The only objective which all parties shared – the adoption of international measures for high seas pollution – was pursued for different motivations and in different ways. Shipowners supported international regulations as long as their foreign competitors would be placed at a similar disadvantage. Government officials did not wish to be embarassed before an influential public, or to allow their ships to be harassed by other countries passing unilateral standards. Environmental groups were concerned about maximizing protection of resources, coastal amenities and wildlife. Although often accused by their protagonists of collectively possessing an 'impractical nature' and held responsible for starting 'an intensive campaign of an offensive character', environmentalists pressed their claims for more efforts to protect the environment – claims which they successfully waged, albeit over a period of time, after palliatives had failed.

2

THE 1926 WASHINGTON CONFERENCE ON OIL POLLUTION

In April 1926, the United States government issued invitations for major maritime states to attend the first international conference on oil pollution control.[38] Care was taken to specify that the delegates should be experts, rather than plenipotentiary diplomats, since the proposed conference was technical and would precede a full-dress diplomatic negotiation for a new treaty on oil pollution. Eventually the 'preliminary conference' became involved in drafting a treaty.

The fact remained, however, that most of those going to the conference had not experienced oil pollution in the same degree as the United States. A number of delegates suspected that they had less to gain from an international agreement than the Americans due to evidence that the problem of oil pollution was most severe in the Gulf of Mexico where crude oil was exported.[39] However, as Hipwood warned in a Board conference with oil and shipping companies prior to the Washington talks, 'If left alone the United States would quite probably enact domestic measures which would be possibly detrimental to British shipping, such as the closing of their ports to ships not fitted with separators.'[40] Therefore, more could be gained by going to the Washington conference and attempting to secure control measures that would be 'reasonable' and 'applied simultaneously by all nations.'[41]

The First International Conference on Oil Pollution

At the invitation of the United States government, twenty-seven delegates representing thirteen major maritime nations met from 8 to 16 June 1926 in Washington, D.C., to discuss the adoption of effective measures against oil pollution of the sea. The following countries were represented: Belgium, Britain, Canada, Denmark, France, Germany, Italy, Japan, the Netherlands, Norway, Spain, Sweden, and the United States. By their presence alone, the delegates indicated the growing international importance of the problem, but the question was whether they could agree on control measures against oil pollution by ships.

In contrast to later oil pollution conferences, which would experience more procedural wranglings and wily legalities, the Washington conference was short and sweet. Charles Kellogg, U.S. Secretary of State, praised it as 'one of the shortest and most expeditious conferences' that he ever knew.[42] Dr. Young recalled that the differences of opinion 'were handled in a good spirit with a view to seeking maximum agreement.'[43]

Ex-Senator Frelinghuysen, who had promoted the conference both in the U.S. and abroad, was elected as its President by acclamation. He placed a global significance to the problem – the practice of oil pollution had become as widespread as oil-fueled ships and oil tankers – and he dedicated the conference to the ideal of total prohibition against operational, routine, discharges by ships.

A Committee on Facts and Causes, with Hipwood as chairman, was organised to consider the extent and effects of the problem in various countries. The first challenge was made by the Dutch and German delegates, both executives of private shipping companies. Mr. F.C. Hannebrink, a Dutch Nautical Inspector, supported by Captain Willi Dreschel of the German shipowners' association, expressed reservations about the need for formal international requirements, because oil pollution, in their view, was not a major problem. Hannebrink introduced scientific evidence – which was in advance of its time – from studies by the Dutch National Institute for Biological Survey tending to show that oil was attacked by bacteria. Captain Dreschel considered that conditions would improve as ships became modernised, and that the areas affected already experienced less pollution. Had the views of the Dutch and German delegates prevailed, the committee might easily have agreed that conditions were improving and did not warrant formal international action. But Hipwood and Frelinghuysen, sensing the doubts, intervened to prevent the conference from being sidetracked. Hipwood insisted that it was 'in every way better that measures of this kind should be dealt with by common agreement and by international action,' than for some countries to take unilateral actions which would discriminate against foreign ships. Frelinghuyen directed the delegates to discuss measures leading towards a treaty.

A Technical Committee was formed to consider the ships which would be regulated and to define the types of oil pollution to be prohibited. They concurred with the American Interdepartmental Committee's report that only discharges of the 'oils of permanency', i.e. crude, fuel, diesel and heavy types of oil, which had a tendency to persist in the marine environment, should be regulated. Lighter grades of oil and petroleum by-products were excluded from the scope of the proposed controls.

Merchant ships and tankers were identified as the main sources of oil pollution of the seas. Land-based sources, through causing similar pollution, were left to the control of individual governments. Government and naval vessels and small craft were exempted as a separate category. Thus any measures were to be applied only to merchant vessels.

An internationally-acceptable definition of pollution had to be agreed upon, for it was an indispensable reference in selecting control policies for ships. Future standards for controlling vessel-source pollution would consist of a level of pollution whereby the Washington conference accommodated the technical capability to achieve such standards, rather than on absolute protection of the marine environment.

Oil pollution experts have often been reduced to perplexity in explaining the basis of the legal definition of oil pollution.[44] The 1926 Washington conference established such a lasting precedent. It based the definition of pollution control, not on absolute purity of discharges from ships but, on a standard of water contamination within the limit of available technology.

Overall, the conference accepted that ships had to dispose of oil-contaminated ballast and cleaning water in one way or other. There were a number of ways to reduce or even to prevent the oily dregs in ballast or cleaning water from being discharged at sea.

A standard of water purity in the oily mixtures corresponding to 95-99% 'clear' was available even then through simple gravity separation onboard the vessel or through the aid of special separators. Ships availing of simple gravity separating techniques or devices would allow the oil to settle on top of the mixture, leaving only that part of the mixture containing from 100-500 parts per million (ppm) traces of the oil which could not be separated completely during the last stages of pumping out the ballast or cleaning water. The latter proportion of oil during the last stages of deballasting or tank cleaning was held to be 'negligible' pollution.

Although the Americans and the British had evidence proving that a higher standard of oil separation was achievable (corresponding to 100 ppm), they did not press, the point. Instead, they acquiesced to a lower standard of water purity (corresponding to 500 ppm), for this was the expedient whereby they thought they might secure a policy of shipboard retention and separation of oily wastes.[45]

Controlling tanker pollution, an especially difficult process due to the large amounts of oily dregs involved, became even more feasible when the Americans attached a rider to the definition of 'negligible' pollution, to the effect that oily discharges which left 'a trail in the ship's wake sufficient to form a film on the surface of the sea, visible to the naked eye in daylight and in clear weather' would fall under the prohibited category. The 'visible sheen' of oil in the ship's wake came from the oily dregs in the last stages of pumping, when the oil and water interface is reached. The oil content in this mixture may be as high as 10,000 ppm, and it roughly corresponds with the oil companies' Load-on-Top discharge, a practice introduced on a wider scale in the 1960s.[46]

Where vessels would now discharge oily washings overboard was discussed in the Committee on Zones, with Dr. Young as chairman. Discussions about zones were attended by some confusion. In the first place, not all the delegates agreed on the necessity for any action that would prohibit discharges on the high seas. The Dutch delegate repeated his contention that oil in the open sea disappeared due to microbiological and other natural processes. As Haanebrink interposed, 'If oil did not disappear, only 6,000 tons are needed to cover the Mediterranean.' Captain Dreschel of Germany was adamant that oil pollution was neither longlasting nor important. 'I have never experienced any oil pollution on the seas,' he told the conference, 'and, as I am in close contact with hundreds and thousands of captains in my long seafaring life, I know they will give you the same answer, that they have never experienced any oil pollution on the high seas.' On another occasion, Dreschel ridiculed the desire for the 'total prohibition' of oil pollution as similar to the American prohibition on liquor sales at the time.

However, the Americans maintained that high seas pollution existed seriously in places, that it tended to persist 'indefinitely' in the marine environment, and that it could be borne by tides, winds and currents to coastal waters, thus aggravating coastal pollution. Dr. Stroop, whose tests for the American Committee tended to prove the persistence of oil at sea, was asked to address the delegates.

Next, the delegates expressed doubts about the international legal implications of oil pollution zones beyond territorial waters. Customary international law at the time admitted only a three-mile width of territorial seas within which the coastal state had the right to regulate ships, exploit resources and protect fisheries. A number of exceptions, for specific functional purpose, such as the prevention of piracy and smuggling, were usually granted to coastal states wishing to regulate activities beyond the limit of territorial seas. But any of these exceptions had been agreed upon only after decades of controversy and excoriating battles with traditional forces, as was to be the case with the fixing of oil pollution zones.

Dr. Young and Hipwood assured the delegates that international law would remain unaltered. Flag states would retain jurisdiction over their vessels, even within the pollution zones of coastal states. If foreign ships discharged prohibited oils within these zones, they would be reported to their flag authority for penalties.

Having been assured that the nature of maritime, jurisdiction would be untouched, the delegates agreed in principle to the fixing of zones. They exchanged ideas on the width of the control zones. The U.S. Shipping Board had earlier advised that 500 miles would in some cases not be too far. Frelinghuysen and Hipwood personally favoured zones of at least 150 miles in width. As a compromise to those delegates who were reluctant to the establishment of zones at all, they agreed to a normal maximum width of 50 nautical miles from the coast, but, in exceptional circumstances such as heavy pollution or irregular coastlines, up to 150 miles might be set provided prior consultations were made with neighbouring countries.

It was obvious to everyone concerned that the Americans were anxious to secure more complete requirements than the zone system. What they really wanted was a requirement for the shipboard retention and separation of oily wastes, but, due to the difficulties of obtaining this system straightaway, they were prepared to accept the establishment of zones during an interim period, after which they had hoped stronger rules would follow. In presenting the decisions of the committee of zones about the fixing of zones and the two-year interim period of adjustment, Dr. Young drew attention to the fact that they had not decided whether the zone system was the extent to which nations would be asked to control pollution of the seas, or whether stricter prohibitions might be recommended as well.

Intense deliberations surrounded this issue, with the

United States. Britain and Canada urging total prohibition by means of stronger rules, and the others either vigorously opposing it or swaying in either direction. Therefore, two schools of thought prevailed. One felt that the oil nuisance would not abate with the zone system alone; the other maintained that the zone system was a sufficient cure for the problem.

In admitting that they were only "guessing at the future,' Hipwood made a strong personal plea for stronger requirements. 'If we have the zone system,' he said, 'we really cannot enforce, it, and we will have this trouble in greater or lesser degree hanging around our necks for an indefinite period, and it is much better for us, and much better for the interests concerned, to settle it definitely once and for all.' But strangely enough, Hipwood waved the flag of total prohibition alone. He assumed the main task of rebuttal and persuasion while the American delegation fell unusually quiet during this crucial period of the debates. Since Hipwood was arguing from a position of weakenss, for he could not singularly command the technical or trade arguments in support of the proposition, whilst the Americans, who had the expertise and clout, did not throw their weight behind the objective they desired, the outcome was decided more or less by default.

A confused ballot on 10th June showed that seven delegations as opposed to three voted in favour of the zone system as only a temporary stopgap measure, with presumbly more complete requirements to follow.[47] But on the following day's crucial debates, support for stronger rules withered under the counter-attack launched by the Dutch and German delegations. Captain Dreschel argued that stronger rules for oil pollution would lead to new equipment onboard ships at a time when they could 'hardly make ends meet'. Haanebrink reminded them that different types of vessels had different requirements, and there was simply no universally applicable oil separator that was at the same time a viable economic proposition.

Evidently, considerations of expediency and economy had greater value for the conference, which rejected the idea of stronger prohibitions than the zone system. When the question was again put to the vote, only the United States, Britain and Canada voted affirmatively in favour of more complete requirements against oil pollution. France abstained. The rest favoured the zone system as the major remedy to be applied against oil pollution.

Although it was not agreed at the start to frame a draft treaty on oil pollution, the delegates recommended that nonetheless they should put their recommendations into a draft treaty. On the next to the last day, Dr. Young and Hipwood drafted the oil pollution convention and final act of the conference in time for the evening's social dinner.[48] The last day of the conference consisted of renewed expressions of co-operation and support for the agreement, and it became the general feeling that the draft convention would receive the necessary support through regular channels without the need for a further conference of plenipotentiaries.

The 1926 Draft Washington Convention

The 1926 draft Washington convention on oil pollution envisaged both a short-term and a long-term answer to the problem of oil pollution by ships. In the first place, the convention provided the possibility of establishing zones on the high seas into which prohibited oil or oily mixtures may not be discharged. The demarcation of these zones was left entirely at the discretion of the coastal state, provided only that they must be no more than 50 miles from the nearest coast, and 150 miles in exceptional circumstances.[49] The coastal state had no authority over foreign vessels in these zones, except the right to inform the flag state of any vessel that made a prohibited discharge within the zone. There was no provision for adjudicating conflicting claims. Secondly, the draft convention tried to reach beyond the zone system as a solution to oil pollution by removing obstacles to the retention by ships of oily waste onboard, even if it did not explicitly require them to do so. The inherent disadvantages in both approaches accounted, in the end, not only for the scorn heaped by environmentalists on the draft convention but also for the delinquency of shipowners and governments in implementing the same.

The system of zones was a difficult issue on which to achieve an agreement or to base one that wanted to control oil pollution by ships. Various countries maintained the suspicion that such zones would be used for purposes other than oil pollution control. Indeed, as we shall see, at least two countries sought to validate their wider limits of territorial seas by declaring them as oil pollution zones. The most telling fault in the zone system, however, was its recognised fallibility in preventing oily discharges from causing coastal pollution. As Hipwood said, 'We know the difficulties of getting evidence within our own three-mile limit. A fortiori what are the difficulties going to be in enforcing it when it comes to a matter of 50 to 150 miles?'[50] Environmentalists were shocked that the conference had even bothered to think of this as a possible solution, be it short-term or long-term. As Miss L. Gardiner, RSPB Secretary, wrote the Foreign Office, 'A regulation purporting to prohibit in general the discharge of oil within a 50-mile zone (or any extended zone) would be ineffective.'[51] But officials were not overly concerned with the actual consequences of the zone system as they were in reaching a compromise on paper – a palliative, as Hipwood told the conference, 'to show people we are really trying to do something.'[52] Those who doubted the persistence of oil at sea were persuaded with difficulty to accept this compromise. Those who doubted the merits of the zone system gambled that the problem would become increasingly worse, so that in time they could re-open the consideration of more effective measures.

Lesser known and appreciated was the fact that the Washington draft convention also encouraged governments to remove every possible obstacle, e.g. customs dues and canal tolls, tonnage measurement, cargo space rules, standing in the way of ships retaining and separating oily wastes onboard. Article VI of the draft treaty noted:[53]

> The respective Governments agree
>
> (a) That no penalty or disability of any kind whatever in the matter of tonnage measurement or payment of dues be incurred

> by any vessel by reason only of the fitting of any device or apparatus for separating oil from water.
> (b) That dues based on tonnage shall not be charged in respect of any space rendered unavailable for cargo by the installation of any device or apparatus for separating oil from water.
> (c) That the term "device or apparatus for separating oil from water," as used in paragraphs (a) and (b) of this Article, shall include any tank or tanks, of reasonable size, used exclusively for receiving waste oil recovered from the device or apparatus, and also for the piping and fittings necessary for its operation.

Without actually compelling shipowners to retain oily wastes onboard, this Article in the draft treaty was intended to encourage such a practice. Even the Dutch expert at the Washington conference conceded that the economies of recycling oil were patently obvious. Evidently, shipowners would accept the need for retaining oily dregs onboard if some practical and commercially viable way could be found to enable them to do so.

The experience of the *Charles Pratt* tanker experiment highlighted various disadvantages in the practice of total prohibition measures. In 1925, when the *Charles Pratt* trial voyage was arranged by the U.S. Bureau of Standards and the Standard Oil Company of New Jersey, both the tanker operator and the terminal arm of the company were prepared to revise their usual practice of completely dumping oily washings overboard. Instead, the tanker successfully separated and retained oily dregs onboard, fitted only with a skimmer. The experiment was so successful that the company fitted two other tankers with similar skimmers and prepared to equip five others. The company found out, however, that by using their tanker for anything but clean ballast, they had forfeited their right to deduct the cargo tonnage used for the retained oil from the official tonnage of the ship. The cargo tanks containing the oil were subject to customs and other tolls, either when the tanker traversed the Panama Canal or came to port with the retained oil in its tanks. For the shipowner to have used or left this apparatus on the tanker, said former company executive James Moss, 'would have placed this owner under substantial competitive disadvantage, and for this reason the project had to be abandoned.'[54]

The 1926 Washington conference had intended to encourage the equipment of ships for onboard retention of oily wastes, yet they failed to carry out this promising initiative. Such a plan, without a legal obligation to do so, could be successful only if various governments agreed to revise their tonnage measurement and customs laws. But it became obvious that the promise outlined in Article VI of the draft convention would fall through. A German transport expert, Mr. P.S. Lahr, suggested that the conference communicate its recommendations on exempting ships equipped for total prohibition to the international conference on load lines in 1930 and to the International Shipping Conference (now the International Chamber of Shipping). He was quickly disabused of the idea by the rest who regarded it 'sufficient for their national experts to keep the matter under examination.'[55] In the diplomatic exchanges accompanying the 1926 draft convention, no indication was given that shipboard retention of oily wastes would be resurrected, at least, for the time being.[56]

Half-way measures often fail to satisfy no one. Environmentalists derided the 1926 draft convention as a total flop. Maritime interests opted for the ineffective zone system which proved a temporary palliative. And governments persisted in the folly that diplomacy would turn their decisions into reality.

Unilateral Measures

The international conference system represents only one of many ways by which oil pollution standards may be improved. To protect their environment, states have also acted on their own by passing unilateral legislation which may anticipate or exceed standards in international agreements. Most governments prefer internationally-agreed control standards, for reasons of uniformity of regulations and international co-operation. For example, the 1926 draft treaty seemed to many major powers as a satisfactory compromise. By contrast, a few countries advocated special measures, which they desired either out of a serious pollution problem or an effort to enlarge their sphere of maritime jurisdiction, or both. Such unilateral acts upset major maritime states which did not want to have oil pollution used as an excuse for changes in international law. Ironically, the isolated acts in the 1920s and 1930s to protect against oil pollution, then abhorred as political heresies to be avoided at all costs, were replicated, with greater public sympathy, in the 1960s and 1970s by major states themselves. Indeed, the international system has now allowed for more flexibility in coastal states' designation of stricter standards and wider zones due to environmental problems, as we shall see in Parts Six and Seven.

Spain (1925)

On 9 September 1925, the Spanish Royal Government issued a circular prohibiting the scattering of oil or oily wastes by ships within a six-mile distance of the Spanish coasts.[57] The British Government immediately objected to the Spanish edict, since it exceeded the three-mile limit.

Despite the potential value of wider anti-pollution zones in principle, such a step was deemed unacceptable to the other interests which Britain wished to promote. British fishing vessels were expected to complain about the curtailment of their fishing grounds. The Admiralty worried that the movement of their warships would be restricted. The prevention of discharge within six miles off the Spanish coast, noted A.A. Hopper of the Board, 'may benefit Spain but is of no benefit to us.'[58] Accordingly, the British Ambassador in Madrid, Sir Horace Rumbold, was instructed to submit an official protect. The Board advised the Foreign Office to challenge the legality of the Spanish regulations without reference to its merits.[59]

Sir Horace submitted the protest, which tactfully drew the attention of Spanish officials to the likelihood of attaining more effective control measures by means of the oil pollution conference in 1926.[60] The British protest, however, did not make the desired impression, and the Spanish edict for a six-mile pollution control zone remained in force.

Moreover, the six-mile limit corresponded with the Spanish position on territorial waters at the 1930 Hague Conference on International Law.

Portugal (1927)

On 29 September 1927, the Portuguese Government in turn decreed that the pouring of oils, gasoline, petroleum and naptha in territorial waters within six miles from the Portuguese coast was to be prohibited.[61] Prior to this new claim, Portugal had various other maritime jurisdiction claims.[62] Hence, the Portuguese concern for oil pollution became another useful rationale to extend their coastal authority beyond the customary three-mile limit.

The British Government's protest drew attention to the results of the 1926 Washington conference and the draft treaty under consideration, but the Portuguese decree was affirmed.[63] Charles Grimshaw of the Board secretly acknowledged 'with satisfaction' Portuguese steps to deal with their oil pollution problem.[64]

The Gulf of Paria (1927-1945)

Trinidad and Tobago, while still a British colony in South America, moved towards unilateral control of oil pollution in the Gulf of Paria on numerous occasions. Here was also discussed for the first time the question of international liability for oil pollution damage arising from offshore oil drilling .

The Gulf of Paria is an inlet of the Atlantic Ocean between Venezuela and Trinidad, about one hundred miles wide and forty miles long. The Gulf has two entrances of about ten miles wide each through the Boca de la Serpiente from the Atlantic and the Boca del Dragon from the Caribbean, and it is a major shipping area for oil and ore exports from the two countries.

On 23 October 1925, Trinidad's Governor-General H.A. Byatt, sent the copy of a proposed colonial bill on oil pollution and asked the Colonial office if they could prohibit discharge completely in the Gulf of Paria.[65] The request was rejected by London immediately. The Foreign Office noted that the Trinidad bill was 'exceedingly precipitate,' at a time when they had just protested the Spanish edict of September 1925, also extending territorial waters, and when they were awaiting the results of the oil pollution conference in Washington. Frederick Adam of the FO Treaty Department ruled that 'there can be no question of regarding this Gulf as a national inlet.'[66] Colonial authorities abandoned the proposed legislation until 1951.[67]

In effect, Westminster officials, were willing to let the colony experience oil pollution in exchange for the false hope of the 1926 draft Washington convention providing the international control measures. Significantly, British officials had information of two other courses of action for the Trinidad problem, which they did not relay to the colony. George Grindle, Colonial Under-Secretary, told Hipwood later that special measures could have been made for the colony since Westminster had the authority to legislate on British ships entering Trinidad.[68] In the first place, Trinidad could have been authorised to close its ports to any ship which it had reason to believe had polluted in the waters of the Gulf or elsewhere. Secondly, a special oil pollution zone of around twenty miles of the colony's surrounding waters could have been designated. These two special measures were kept secret, however, due to the impending conference in Washington and the legal anathema of the proposed measures.

Meanwhile, the commercial potential of the Gulf became more important than the protection of its waters from oil pollution. After favourable reports of rich oil deposits in the Gulf of Paria seabed, the Colonial Office in January 1931 suggested the conclusion of a treaty with Venezuela dividing the control of these resources.[69] Legal advice held that claims to the seabed and subsoil were sustainable, and that a treaty between the two countries for these resources had to be settled.

Trinidad's Mines Department saw the need to consider the environmental consequences of offshore oil drilling. Since 1906, the United States had pioneered the drilling of oil in the Gulf of Mexico, and, although a certain amount of spillage emanated from these activities, the experts found no precedent on which to base rules guarding against such pollution, except general requirements to exercise care and to use modern techniques in drilling.[70] Technical experts in Trinidad prepared the drilling regulations with due regard for pollution.

At one stage, they even considered taking early regulations to establish strict legal and financial liability for oil pollution damage due to offshore drilling. R.S. McKilligin, head of the Trinidad Mines Department, inserted a clause in the proposed drilling regulations providing for a maximum fine of Trinidad $2,500 for a first offence, and Trinidad $5,000 for subsequent offences, in cases of spills resulting in damage to fishing or coastal amenities in the colony. The suggestion to compensate for pollution damage was based on strict liability or without having to prove fault, negligence or the lack of foreseeability on the part of the drilling rig operator. However, the Petroleum Department in London vetoed the idea of strict liability and a specific fine for pollution due to the nature of accidents and operational discharges in offshore rigs. Instead, they endorsed only a moral obligation against pollution.[71] Admiralty experts similarly rejected a Colonial Office request to protect the fishing industry, because they anticipated that 'some pollution of fisheries' was unavoidable.[72]

Thus, Article 5(2) of the Submarine Areas of the Gulf of Paria (Annexation) Orders-in-Council, issued on 6 August 1942, contained the following general requirement against oil pollution:[73]

> 5. The Governor of the said Colony shall, as soon as may be possible after the date of this Order, make regulations to ensure . . .
>
> (2) that all practicable measures shall be taken to prevent the exploitation of any of the said submarine areas from causing

> the pollution of the coastal waters by oil, mud, or any other fluid or substance calculated to contaminate the seawater or shoreline.

The drilling regulations issued in 1945 also followed the same pattern.[74]

During negotiations for the treaty with Venezuela, R.N. Quirk, the new head of the Trinidad Petroleum Department, advocated inserting a similar clause in the proposed treaty.[75] Venezuelan authorities accepted the clause, and the final treaty between Britain (for the colony of Trinidad) and Venezuela, which was signed on 26 February 1942, contained the following Article VII:[76]

> Each of the High Contracting Parties shall take all practicable measures to prevent the exploitation of any submarine areas claimed or occupied by him in the Gulf from causing pollution of the territorial waters of the other by oil, mud, or any other fluid or substance liable to contaminate the navigable waters or the foreshores and shall concert with the other to make the said measures as effective as possible.

The Trinidad drilling regulations became a model for coastal licences in other colonies. The 1945 Bahamas Petroleum Act and the 1949 British Honduras oil drilling regulations also contained similar environmental clauses.[77] At present, pollution damage due to offshore drilling activities in the North-West Atlantic and the North Sea have been provided for in two international arrangements – the 1976 Convention by governments and the 1974 voluntary liability agreement by oil companies.[78]

Diplomatic Treadmill of the 1926 Draft Convention

Although American officials played the major role in negotiations for the 1926 draft Washington convention, British officials also often became actively involved. Ultimately, the Americans abandoned the draft treaty, and it was left up to the British officials to cobble together another convention.

The 1926 draft Washington convention suffered from external political misfortunes and from a weak legal format. A technical treaty on this topic received a very low priority in the work of officials in the major powers, except for a few enthusiasts like Dr. Young and Hipwood whose brain-child the draft treaty had been. As most American officials turned complacent and British functionaries refined the text of the draft treaty, several maritime states stiffened their opposition to the very idea of imposing more international obligations on their ships. The draft Washington convention seemed ill-fated in the face of other developments in the maritime industry, the decrease of pollution, and the onset of the Great Depression in 1929-1934, all of which impressed greater caution against radical changes in the oil and shipping trade, due only to oil pollution control.

However, as we shall see, it was not inconceivable that, had the Americans and the British exerted consistent efforts and put the weight of their influence behind their good intentions, the draft treaty could have been formalised. We shall now examine two case studies to see how intramurals within ministerial departments and among government ministries, as well as inter-governmental squabbles, though seemingly unrelated to the problem on hand, caused long delays and ultimately proved fatal to the draft convention. Presently, those who wonder at the length of time in which it takes technical maritime treaties to enter into force, up to a decade in the case of the 1973 Marine Pollution Convention, may benefit from the historical insights provided by these secret exchanges of government officials.

Britain and the Dominions

Even before the Washington conference, the political relationship within the British Empire was undergoing great changes, a process that would lead to the transition from Empire to 'Commonwealth'.[79] The controversies included diplomatic questions whereby states symbolize their identity, independence, and equality. Different strategies operated in different countries, with Canada, the Irish Free State and the

Union of South Africa particularly eager to assert their independence from Britain. These diplomatic imbroglios delayed the negotiation of the 1926 draft convention. As the political balance between Britain and the Dominions changed, ironically enough, officials became even more obsessed about the symbols of sovereignty – flags, forms of treaties, the use of legal terms, etc. This obsession with appearances would probably have baffled outsiders, but purists played the game with such seriousness as to signal that deeper motivations lay in store.

On 15 October 1926, the British Cabinet approved a recommendation that in future technical conferences, Britsh officials should endeavour to ascertain an equal position for the Dominions and an unambiguous obligation for the colonies and protectorates. The Foreign Office Treaty Department, which had taken charge of the documentation for the 1926 draft convention on oil pollution, urged that it be made the seminal agreement to confirm the new policy. Thus, Britain's new attitude to treaty obligations as regards Dominion involvement would be tested by using the 1926 draft oil pollution treaty as a point of reference.[30]

Dominion merchant ships had customarily flown the 'Red Ensign' differenced in each Dominion only by a separate coat of arms. On the desire of the Irish Free State and the Union of South Africa to have different flags for each Dominion's vessels, Adam of the FO Treaty Department perceived a leverage point to deter this move. He noted that the 1926 draft oil pollution treaty might be 'a convenient peg on which to hang' the separate flags controversy. These Dominions might be less eager to fly different colours if they realized their ships would be under penalty for oil pollution offences, and an argument against separate ensigns, if couched in those terms, might make the Dominions reconsider their demand. 'In this way,' Adam wrote, 'the "strange devices" of General Hertzog and the Irish Free State might disappear from the seas – which would be a good beginning.'[81]

After the 1926 Imperial Conference had arrived at the definitive interpretation of the political relationship between Britain and the Dominions, granting the latter equality in all the functions of statehood (except certain areas of diplomacy and defence), the FO legal advisers once again suggested that the 1926 draft oil pollution convention be made the occasion to introduce the new treaty-making arrangement between Britain and the Dominions. This involved deliberations on whether the treaty should be in the form of a Heads of State treaty or an inter-governmental agreement, and secondly, whether the *inter se* doctrine applied to the draft convention.

The draft convention was considered too imprecise and informal. It was drafted, after all by Dr. Young, an economist and not well-versed on the legal intricacies of the British Commonwealth. While the complex nature of Britain's new relations with the Dominions became interwoven with the fate and form of the 1926 draft pollution treaty, the United States and other countries, which made no distinction between types of treaties and had no imperial axes to grind, were kept waiting for the British reply.

For several years, various ministries in the British government, consulting with the Dominions abroad, proposed amendments to the 1926 draft treaty. These amendments included the form of the treaty and the proposed inclusion of an *inter se* clause, a specific declaration stating the application of the treaty to Britain and the participating Dominions.[83] The process of collating these amendments from intra-departmental minutes, inter-departmental exchanges, and international correspondence proved so slow and frustrating that a Foreign Office official stated at one stage, 'The Dominions Office must manage their own squabbles.'[84]

It was finally decided to defer all British amendments since the draft today showed no signs of coming to life, and British officials did not wish to add to other questions at issue with the Irish Free State and the Union of South Africa.

The Central Agency

Another delay was caused by the American suggestion to change the location of the proposed central agency from

Washington to London.[85] Article VII of the original draft convention had stipulated that a central agency in Washington would receive and circulate data on the oil pollution zones. This, the precursor of the present-day International Maritime Organization of the United Nations, was expected to undertake a number of other functions, such as the collation of statistical and technical information, reports of prosecution and experience with the zone system, the provision of a library service for the convenient reference and guidance of states. In the process of deciding on the site of the central agency, the states concerned, and ministries within the government, deliberated kindred political issues which had little if any to do with the technical functions of the agency.

The ostensible reason given by State Department officials was that they were concerned lest Senate ratification of the treaty be delayed and the agreement enter into force without the United States being able to arrange the necessary work. Furthermore, London was a more central location, and the British had an outstanding experience in maritime questions. Dr. Young recalled that American shipping interests had advocated a number of strong measures to discriminate against foreign shipping, and the State Department wanted the central agency away from these reactionary interests.[86]

The Board at first contemplated accepting the American offer. As A.A. Hopper noted, they had 'the opportunity and had better make the most of it' because there 'may be certain practical advantages to be gained by having the agency in London.'[87] However, the Board and the Foreign Office suggested declining the American offer, mainly in order to involve the United States in maritime problems and because the central agency would not harm British interests if it were sited in Washington.

The Admiralty and the Colonial Office, however, had entirely different approaches. The Admiralty insisted that the central agency had better be located in London due to their mistrust of American impartiality and competence. The Royal Navy Hydrographer, Captain H.P. Douglas, noted:

'the agency, if located in America, would probably become in everything but name, a centre of propaganda run for the benefit of American shipping interests.' Why should Britain give the United States a potential new weapon against British shipping interests 'when we can keep it ourselves – not necessarily to use against them, but to make sure that no one uses it against us? '[88] The Colonial Office thought the central agency was best placed under the aegis of the League of Nations.[89]

In April 1927, Dr. Young attended a meeting in London with British officials on the proposed draft convention.[90] At this meeting, it was agreed that while the United States canvassed other governments on their position vis-a-vis the draft convention. Britain would approach them on the question of the location of the central agency. After receipt of the replies to these inquiries, the United States would consider the British amendments to the draft treaty and whether a further conference of plenipotentiaries was necessary for the signing of the oil pollution agreement.

Before this could be fully realized, another confusion and delay arose. In June 1927, the State Department asked that the question of the central agency be postponed until the draft convention itself had been accepted by the other governments.[91] Six months later, however, the Americans admitted that there had been a misunderstanding in this postponement, and Ray Atherton, the First Secretary in the U.S. Embassy in London, was asked to urge the British Government to forge ahead with their own inquiries on the site of the central agency.[92]

At this stage, American reaction was encouraging, even though the State Department contemplated the necessity of another conference to formalise the agreement. More formidable barriers, however, were forthcoming.

The Decline of the 1926 Draft Washington Convention

By 1927, the United States and Britain had co-ordinated two separate but related inquiries, with the United States canvassing maritime nations on their views regarding the draft

Washington convention on oil pollution, and Britain collating the views of nine countries present at the Washington conference on the establishment of the central agency in a site other than Washington. A year later, the policies of the other governments were made known, and it would only have remained for the United States to summon a diplomatic conference to conclude the agreement. However, it was not long after this period when things went awry.

In February 1928, Ronald Campbell of the British Embassy in Washington was instructed to inquire on the progress of the American démarche. State Department officials either disclaimed interest in the subject of oil pollution 'or showed boredom with the whole subject,' according to his report, except for Dr. Young who evidently maintained his enthusiasm for his brainchild.[93] Dr. Young assured Campbell that positive results were being obtained and that no delay was foreseen. At this stage, Dr. Young felt there were no major problems, even as regards the British amendments to the draft convention, and that their delegates could meet again in Washington to formally sign the treaty.[94]

Thus, by the end of 1928 the hopes for securing an oil pollution treaty seemed bright. Both the United States and Britain had sufficient information from replies to their inquiries on which to base the next course of action, which, under the circumstances, would have meant another international conference to conclude the treaty. Major maritime nations favoured the oil pollution agreement. Belgium, Canada, Denmark, France, Norway, Spain and Sweden agreed in principle with the draft and were prepared to sign immediately. The Dutch government, whose representative at the 1926 Washington conference strongly objected to an oil pollution treaty, now indicated their approval of the same.

Controlling oil pollution, however, depended on a host of factors outside the controlof the negotiators, despite their personal interest in the problem. Overall, the voluntary measures implemented by major shipowners (which we shall discuss later) and the effect of various national laws contributed to a decrease of pollution. The onset of the Great Depression prevailed against imposing costly new procedures on the oil and shipping trade. And finally, there arose several problems of a diplomatic nature that had to be nursed through their difficult periods if the United States and Britain wished to persist in obtaining an oil pollution treaty.

Portugal, whose opinion was solicited even if they did not attend the 1926 conference, was willing to sign the treaty but wanted wider zone authority.[95] The British were aware of Portuguese intentions to extend their territorial waters by using oil pollution zones as another excuse, but their reply was allowed to pass without protest or inquiry due to more formidable barriers presented by other maritime states.

While the Japanese Government did not object to the siting of the central agency in Washington, they did not reply to the feasibility of the proposed convention at all. This created some uncertainty.[96]

The Royal Italian Government were determined to urge a second international conference to re-evaluate the question of high seas control of oil pollution. The Italians refused to accept a draft treaty without a 'fuller examination' of its details, but did not specify their particular doubts or objections.[97]

German opposition was more formidable, and their objections to the treaty hardened as they sensed a decreasing interest on the part of the other countries. German maritime interests had little difficulty in persuading their government to keep the oil pollution treaty in abeyance. The German shipowners association told the International Shipping Conference in December 1927 that they would not participate in a voluntary system of oil pollution zones. In 1928, the Hamburg shipowners noted that the German Government had decided that ratification of the draft convention was 'impractical' at a time when the question of territorial waters had not been settled. The report also noted with satisfaction that international complaints were diminishing.[98]

It appears that the Reich Foreign Ministry conveniently brought up questions of maritime jurisdiction over ships as an excuse to delay and to thwart the draft convention. In reply to British inquiries, German functionaries gave different explanations at different times – e.g. Germany was apprehensive lest the treaty authorise an extension of territorial waters; Germany believed that the establishment of oil pollution zones should be examined before, rather than at, the second conference; and, finally, they said that the United States was preparing to abandon the draft convention anyway. The British were totally perplexed by the German replies and wanted to know if deeper motivations lurked beneath their misgivings. A Foreign Office official could only surmise that they did not know, except for 'German reluctance to agree to anything, no matter how reasonable or advantageous, except in return for some probably totally unrelated concession.'[99]

The Germans managed to prolong the diplomatic exchanges for nearly ten months by their dilatory tactics so that the Foreign Office became furious at one stage. 'The Germans have discovered or think they have discovered another stalking horse behind which they hope still longer to evade the agreement,' wrote C. E. Steel.[100]

Dr. Young, however, did not feel that Germany would present substantial problems to the conclusion of the agreement.[101] Unfortunately, by the end of 1928, Dr. Young vacated his position in the State Department to advise the Chinese Government on financial matters, and, with his departure, American officials gave little further attention to the draft convention.

Good results had been obtained from the 1924 Act and from the subsequent voluntary measures undertaken by American oil and shipping interests. Secondly, follow-up experiments conducted by Dr. Stroop of the Bureau of Standards failed to support the theory that oil in the marine environment persisted indefinitely, as American officials had earlier believed.[102] It was inevitable for American officials to believe that, if high seas pollution did not persist indefinitely but was in fact susceptible to natural degraduation over a period of time, and their own coasts were sufficiently protected by domestic measures, then international agreement on high seas pollution was completely unnecessary. Moreover, with the departure of Dr. Young for China, the guiding spirit of the 1926 draft convention had been lost.

The British, after being informed of the results of Dr. Stroop's experiments, despaired of securing the draft today, even though they had scientific studies tending to disprove these findings.[103] Some British officials tried to maintain the momentum until 1930, but these were instances of personal initiative or a reflex reaction to pressure by domestic environmental groups.

Ronald Campbell in Washington tried on several occasions to inject interest among other diplomats on the oil pollution negotiations. Although his efforts went largely unappreciated by some colleagues in London, at least Campbell's initiatives enlightened the Foreign Office on the status of play. [104]

Anxious about parliamentary inquiries on the subject, Board officials urged the Foreign Office to ascertain American position on the draft convention. At one stage, they were informed that, although Congress had voted the appropriations for the oil pollution conference, the State Department had no interest in calling for one. [105] Dr. Stanley Hornbeck, the successor of Dr. Young in the State Department, categorically replied that under no circumstancs would the United States pursue the adoption of the draft oil pollution convention.[106]

In December 1929, Sir Henry Payne of the Board called for a meeting with British oil and shipping interests to inform them of the stalemate in the oil pollution negotiations. They discussed the possibility of a modified zone system for the English Channel or the North Sea.[107] However, before proceeding with the idea of a limited convention, or picking up the threads of the negotiations for the 1926 draft treaty, the Board decided to make a new general inquiry on the problem of oil pollution round the coast of Britain. On the basis of this report, it was decided to follow the example of the Americans and drop the idea of securing international agreement on oil pollution control. [108]

The Wall Street collapse in October 1929 began a period of economic depression and political uncertainty in world affairs. The Great Depression did not reach Europe until the

1930s, but world trade suffered a drastic slump, and shipping was similarly affected. Against this background, the protection of the environment became the farthest concern in the minds of commercial leaders who were desperately trying to survive economic ruin. Moreover, the decrease of oil pollution in coastal waters, due partly to national legislation, to voluntary measures taken by port and maritime interests, and to the decrease in shipping activity in general, made international action less urgent to policymakers. Finally, the Americans, after the departure of Dr. Young from the State Department, abandoned the project. The British, who could have taken the initiative, were only forced to do so after a period of time and in another forum.

Raking the Ashes

British environmentalists never really accepted the 1926 draft convention as the best means of controlling oil pollution, and they mounted a vigorous campaign for more complete requirements. Maritime interests opposed any measures that would touch on the question of new equipment for ships or port facilities if foreign shipping would be excused by their government from taking similar action. In the latter's view, a unilateral undertaking to compel British ships to equip for total prohibition was impractical and financial suicide at a time of economic depression. British officials were generally inclined to swing to the side of the maritime interests, but they could not continue to ignore domestic agitation and political uncertainty about the actions of other countries.

The Royal Society for the Protection of Birds (RSPB) and the Royal Society for the Prevention of Cruelty to Animals (RSPCA) enlisted the support of influential members of Parliament and other prominent leaders in British society to the cause of compelling ships to equip for environmental measures. Sir Philip Cunliffe-Lister, President of the Board, faced numerous questions in parliament, to which he could only reply that the Government's policy was to urge the adoption of zones in the 1926 draft convention.

To a letter in 1927 of several MPs, Sir Philip flatly stated that he was not prepared to compel British ships to fit separators.[109]

In 1928, the campaign peaked for domestic legislation on shipboard separation of oil. The RSPCA circulated a pamphlet (see Fig. 1) illustrating the damage of oil pollution to birds in order to attract public support. According to this publication, oil pollution, if allowed to continue, would mean that 'future generations will deplore the uncivilised actions of nations who did not insist that efficiency in engineering requirements to preserve the beauties of the world, should keep pace with the oil-burning tendencies of the age.'[110]

In November 1928, Sir Cooper Rawson, M.P. for Brighton and Howe, drafted a proposed bill compelling British ships to equip for total prohibition and presided over a much-publicised caucus at the House of Commons. Due to the defeat of the Conservative Party in the 1929 elections, however, the bill was not pursued with until 1931.

Meanwhile, a public row erupted between the environmentalists and shipping interests on the costs and benefits of equipment for retention of oily wastes. Various statis-

Figure 1. RSPCA Pamphlet on Oil Pollution in 1928
Source: PRO ADM 116/2624

tics were flaunted on equipment costs (ranging from £250-£350 thousand) on benefits (savings of about 500,000 barrels or 70,000 tons of oil). *The Financial Times* estimated that 500,000 barrels of oil (worth £250,000) were annually tipped into the sea. [111]

When it became clear that the United States had abandoned the draft convention, Sir Henry Payne of the Board undertook certain initiatives on the advice of Hipwood that 'there may be trouble' with their domestic public unless the Board showed something was being done. [112] Meanwhile, it appears that Hipwood also secretly advised British environmentalists to lobby the Board to bring the question before the League of Nations. In 1929, the RSPB noted the 'melancholy deadlock' in the diplomatic negotiations and suggested action through the League, a proposal which received co-ordinated and warm support from the various League of Nations unions and local groups of the National Woman Citizens Associations, which were presented to the Board or the Foreign Office in mimeographed form or in formal letters. [113]

By 1931, both the RSPB and the RSPCA combined forces to support Sir Cooper Rawson's new bill for total prohibition, which was re-introduced as a private member's bill. In the throes of the depression and the difficulties of British shipbuilding and shipping, Sir Cooper nonetheless aroused cheers and applause in the House when he said, 'we have reduced our Navy very materially, but we are still the largest maritime power in the world, and although Britannia may not rule the waves, there is no reason why she should not keep them clean, and set an example to other countries.[114] Without the support of the Labour Government however, the bill was withdrawn before the second reading.

Shipping interests had friends in high places as well. Their spokesman in the House of Lords, Lord Strabolgi, was briefed by a letter from the Chamber of Shipping on how to answer parliamentary questions. Privately, the Chamber complained that the environmentalists were not entirely altruistic: 'the chief fault is believed to lie with the Society for the Protection of Animals [sic], who apparently wish to rival the specialised Society for the Protection of Birds, and both are known to have been encouraged – if not indeed financially supported – by persons interested in the commercial exploitation of separators.' [115]

It seems that good causes never die, even if they momentarily fade away. In 1933, British officials changed their minds, and, largely in response to domestic public opinion, reassessed their policy on the oil pollution issue.

In July 1933, the Board President and the Minister of Agriculture and Fisheries were asked to receive a deputation and to consider a petition signed by 530 members of both Houses of Parliament. The deputation and petition requested the government to form a Joint Select Committee to hear evidence and make specific recommendations on further control of oil pollution.[116] Despite good reports on the decrease of oil pollution round the British coast, the Board and the Foreign Office became uneasy that such growing public support for further measures might not so easily be parried as before. They agreed that, if nothing else but to show some action was being taken, they should revive international consideration of oil pollution control, this time through the League of Nations. [117]

Before taking the matter to the League, however, they wanted the support of other nations, in particular that of the United States, Germany, Italy and Japan, the countries which presented problems before. There was another reason for soliciting the views of these countries, as C.A. Evelyn Shuckburgh of the Foreign Office noted: Britain was still free to do nothing if their replies were unfavourable and would 'be in a strong position in the public eye, having pressed for action, and been prevented by the coolness of others.'[118] J.W. Nicholls of the same department reasoned that if the British Government were to find it difficult to raise the question at Geneva, public opinion might be convinced that Britain had been hindered 'in their humane endeavours by obstructive foreigners.' [119]

Before Parliament considered new measures on oil pollution, the Board and the Foreign Office hurriedly enlisted the support of British environmental groups and their own personnel to approach similarly-minded groups in the United States and other maritime countries. Leander McCormick-Goodhart, honorary commercial secretary of the British Embassy in Washington, who was affiliated with American conservation societies, suggested that British societies should approach their American colleagues, albeit secretly so that the British Government might not lose face in case of an official rebuff. Apparently, these private initiative proved successful.[120]

There were other reasons for reopening the international control of oil pollution. It was not inconceivable that Britain or other countries might be forced, by domestic pressure or the menace of pollution, to take unilateral steps detrimental to oil and shipping interests. As previously seen, Portugal used oil pollution as the occasion to affirm its extended claim to territorial waters. At the 1930 Hague conference on international law, the American delegate also cited the possible effectiveness of coastal state jurisdiction over oil pollution zones beyond the limit of territorial seas.[121] British officials wished to ward off any such measures, which they interpreted as a threat to their shipping interests. As Geoffrey Thompson of the FO noted: 'We would welcome international action . . . to still the protests of the bird societies and to safeguard our mercantile marine from the risk of suddenly being subjected to arbitrary and possibly ill-advised or unworkable rules which Portugal or some other country might seek to impose upon foreign shipping.[122]

The League of Nations therefore had its uses. To the world, Britain could pose as the leader in controlling oil pollution control, and to a domestic audience, Britain could sweeten the grim reality of a calculated strategy to compromise diverging claims at home.

3

THE LEAGUE OF NATIONS AND OIL POLLUTION

Before elevating the problem to the League of Nations, Britain faced three obstacles which had to be cleared. These were as follows: firstly, the question of selecting the best possible measure to control oil pollution; secondly, how to present government policy to achieve maximum success with all affected interests; and, finally, whether the government wanted to take up this problem yet again, knowing the attendant difficulties involved.

When environmental agitation in Britain reached a new height in 1933-1934, the Foreign Office briefly toyed with the idea of rushing the question to the League of Nations at the earliest possible moment, even without the express support of the four maritime nations they had asked. But the Board of Trade cautioned against moving without the firm assurances of these key states. Apparently, other countries had become as anxious to discuss the problem, as evidenced by an inquiry from the Danish government in October 1933.[123]

Whatever the FO thought about the delay, they deferred to the Board on matters of maritime policy. Unfortunately, at this time the Board seemed to suffer from a paralysis of will due to internal re-organization. When Hipwood retired in 1932, few officials showed the same level of personal interest to take initiatives on oil pollution control. It was not until after Charles H. Grimshaw, Assistant Secretary, had settled in his new post that the Board regained some of its former stride. However, Grimshaw's field of manoeuvre was more restricted than Hipwood's dominance, and, with the passage of time, the novelty of the problem had worn out and unfavourable attitudes had hardened. Moreover, it appears that Grimshaw's pet project was the improvement of coastal erosion defences round the British coast, rather than oil pollution control.

The July 1933 petition by 530 members of both Houses of Parliament and the preparation for a Joint Select Committee to review government policy on oil pollution control worried the Board. The Select Committee might exert pressure to compel British ships to equip for retention of oily wastes onboard, even though the Board believed that this measure would place their ships at a considerable disadvantage.

By early 1934, replies to the British inquiry to bring the problem before the League had been received from three countries, with the exception of only Germany. The United States, Italy and Japan expressed their reluctance to impose additional oil pollution control measures on their shipping at a time of economic depression, but none of them rejected outright the idea of some co-operation within the framework of the League.[124] It was now up to Board to decide on the next step.

On 3rd February 1934, the Board decided that Britain should bring the problem of oil pollution to the League of Nations. In their letter to the FO, it was noted that 'public outcry in this country will not be satisfied until H.M. Government have done all that is possible'. Moreover, in this manner, it was thought that British officials could stave off the possibility of domestic inquiries and stricter legislation by Parliament.[125]

The irony was that British officials also realised that, even as matters were advancing in the direction which environmentalists wished them to take, their initiatives might not yield a happy ending for everyone concerned. There was little hope of securing a more effective solution than the system of zones, if international opinion was favourable at all to any form of oil pollution control. But the zone system, an acknowledged palliative, was held in disdain by the very group of critics they wished to placate.

Before airing the issue at Geneva, it was considered prudent to settle the question of how to present policy in such a way as to make it seem that Britain was reasonably inclined to take strong measures if and when the other countries wished the same. Whether Britain would espouse the cause of the environmentalists or provide the lead for total retention of oily wastes was regarded as a sensitive question and was referred to the Board President finally to decide, since the Board ranks appeared reluctant to advise on a forthright line of action against what consequence may arise. The Board President, Sir Walter Runciman, was given a delicately-balanced memorandum, written by Grimshaw in 1934, outlining the various alternative modes of oil pollution control, the effectiveness of each, and their consequences upon maritime interests, public opinion and international affairs. Runciman was specifically asked to choose whether Britain should press for a more effective system than zones – the compulsory retention of oily wastes either onboard ships or the provision of port facilities. There was no doubt, according to the Board memorandum, that 'the most effective of remedies for the evil of pollution would be international agreement providing for the compulsory use of separators on certain vessels.' The system of zones, 'even if their width should be increased beyond 50 to 150 miles, would be only a palliative and not so satisfactory a remedy as the compulsory fitting of separators on some ships and the provision of port reception facilities.'[126] However, not all the facts were presented to the Board President. The Admiralty had information about a special process which was unfortunately withheld at the time. This secret process, developed by the Admiralty's Military Branch and used since 1929 on military ships, was readily applicable to commercial vessels.[127] By the time the Admiralty offered to distribute the patent more widely in 1936, it was too late, since policy had been determined by Britain and, later, finalised by the League of Nations Committee of Experts.

The Board President was reminded that British public opinion had expected the government to press for the best possible means of controlling oil pollution as the basis for international agreement, and not just perform an exercise in futility.[128] Essentially, Britain was caught between two powerful lobbies.

Had attitudes been more enlightened and times less harsh, the final decision probably might have inclined toward the more effective, but, at the same time quite radical measures. The Board realized only too well that they could not expect their oil and shipping interests to shoulder new and expensive investments when the latter, in forma pauperis, were suing the government for subsidies to offset their economic losses. As one FO official commented sadly on the matter, 'Not even this country is going to have its ships especially designed for the sake of birds, however harrowing their death by oil may be.'[129]

The British delegates to the League therefore were instructed to speak publicly in favour of any recommendation to control oil pollution, but to opt at the moment of decision for the zone system. In such a way, officials hoped that Britain would still be seen as having brought the problem up for international action, but would throw the onus of rejecting the more effective measures onto other countries. According to the secret instructions to British delegates, 'it will be very easy to avoid this onus since almost all shipping countries are known to oppose it.' The policy of the British Government was to join the experts at the League of Nations, quickly ascertain their views, and then to report back that 'unfortunately' there was 'no possibility of securing agreement' on total retention methods. Secondly, they would press for acceptance of a system of prohibited zones of 50 to 100 miles (according to the FO) or even 200 to 250 miles, with possibly greater widths for exceptional circumstances (according to the Board), accompanied by a strong recommendation for the provision of port reception facilities at free or nominal charge. If the enlarged zones and the recommendation for port facilities failed, too, then their fallback position was for any type of zone system, which was 'better than no convention at all, though it was not effective.' [130]

The League Assembly as an Agitation Chamber

By 15 May 1934, governments who attended the 1926 Washington conference had been informed that Britain would raise the question of oil pollution at the next meeting of the League of Nations Assembly in September. It was necessary to follow certain set procedures in League protocol, a delay which made some Board officials anxious. Ralph Stevenson of the FO informed Robert Haas of the League's Advisory and Technical Committee on 19 July 1934 that Britain wished to bring the oil pollution problem to the League. He was advised to send an official letter to the League Secretary-General, to be inserted in the annual report of the League's Communications and Transport Organisation (CTO), the agency within the League dealing with maritime affairs. [131] At the appropriate time, the British delegate at the committee considering this report would draw attention to the relevant passage and move for the consideration by the League of the oil pollution problem.

The Board felt dismayed at the cumbersome procedure of the League. Grimshaw felt that the process of shunting the question from the League Assembly to the Council by way of the CTO, 'with its ten or eleven stages' was a long drawn-out agony. [132] But he agreed to it when informed that the only alternative was for Britain to convene an international conference, which was out of the question.

Another minor tempest erupted between the Foreign Office and the Board over the remarks by the British delegate who introduced the subject at the League. Sir Geoffrey H. Shakespeare, Parliamentary Under-Secretary at the Ministry of Health, represented Britain at the League Assembly's Second Committee, and accordingly, he was instructed to draw attention to that part of the CTO's annual report dealing with the need to secure international control of oil pollution. Since British government officials were most anxious to appear well before their own domestic public, the speech which Sir Geoffrey was to deliver had been carefully prepared and scrutinized by various ministries. With many minds working on a draft, it was inevitable that some ministries insisted on changes, in particular, a deletion on damage to fish and fisheries, for which it was said that neither the Board nor the Ministry of Agriculture and Fisheries could find evidence.[133] The FO thought the appeal of the speech had been considerably weakened by this particular change.[134] When the British delegation to the League met Foreign Secretary Sir John Simon, however, the latter told them to consider oil pollution as a 'scandal' which called for immediate international action and urged them to make the strongest possible case for it. Thus, when the time came for Sir Geoffrey to deliver his prepared speech before the League, he was inspired to exceed the bounds of the draft and to elaborate on the damage oil pollution can inflict on birds, beaches – and fish. The Board adviser to the delegation was upset and personally cautioned Sir Geoffrey about the matter, but the latter replied that he knew what he was doing as he had attended Professor Gardiner's zoology lectures at Cambridge and heard arguments about oil damage to fish.[135] In London, the Board complained to the Foreign Office, but they were informed that Sir Geoffrey's reference to fish and fisheries damage 'restored to the speech some semblance of conviction.'[136]

Thus, on what appeared to be sheer determination and rhetoric alone, the British Government had managed to convince the other League delegates to vote in favour of a resolution endorsing the case for international control of oil pollution. An ad hoc committee of experts was appointed to examine the problem further.[137]

The League Committee of Experts

Delegates from six countries (Britain, Denmark, France, Italy, Japan, and the United States), comprising the newly-appointed League of Nations Committee of Experts on oil pollution, met for the first time at Geneva from 19 to 23 November 1934, with Charles H. Grimshaw as Chairman.[138] Most of the experts present were skeptical of the value of a formal international convention on oil pollution control, since they believed that the problem was less severe than a decade ago. However, a few important maritime states were persuaded to the idea of resolving the matter once and for all, if only to placate their environmental critics at home and to guard against the possibility of unilateral measures being forced on any of them or being taken by other countries experiencing pollution. Much to the relief of British officials, the Committee rejected the compulsory retention of oily wastes onboard ships by means of separators. And, by unanimous decision, the experts agreed that the course most likely to commend itself to their respective countries was the system of oil pollution zones. Anticipating this outcome, Grimshaw had brought along a copy of an amended British draft of the 1926 Washington convention, and the other delegates agreed to use this as the basis for future discussions.[139]

The Committee also prepared a questionnaire for various countries to render their reports on the extent of pollution and their adoption of various anti-pollution measures. The international survey was expected to show that 'the zone system was the only measure which in present conditions would meet international support.'[140]

The League experts were quite anxious to keep their tactics a secret from the public. They fully expected the questionnaire results to support a case for a convention primarily based on the zone system. However, they were in for a surprise.

The League Questionnaire on Oil Pollution

On 23 January 1935 the League Secretary-General circulated the international questionnaire on oil pollution to seventy countries, and the latter were asked to submit their replies in six months time.[141] Prepared by the Committee of Experts under Grimshaw's lead, the questionnaire solicited information on the effects of oil pollution; opinions on the compulsory fitting of separators on new and existing ships; types of separators used and supply of port reception facilities; scientific or other evidence on the behaviour of oil at sea; and any other comments by government. In many ways, this survey was more extensive than the one made by the American Interdepartmental Committee in 1925.

Thirty-four countries responded to the League questionnaire, and about 60% said there was little or no harmful consequences from oil pollution in their area, or otherwise they gave no evaluations. Slightly over 10% of the respondents described moderate to serious damage, especially to birds, and these countries had extensive coastal and shipping interests. The rest, or 30%, gave no particular information on the damage of oil pollution. Twelve important trading countries ascribed serious damage resulting from the pollution, while twenty-two countries reported it to be only an occasional or negligible problem.

Thus, on the basis of these replies, the League Committee of Experts concluded that pollution was 'serious enough' to warrant international action. British shipowners, however, came to the opposite conclusion and protested that the same survey showed the problem was less urgent than a decade ago.[142]

The remaining questions in the survey were intended to attract opinion and reports on the various means of control – from the zone system to the fitting of separators and the provision of port reception facilities. The respondents were not asked to express their opinion on whether oil pollution zones in the high seas would be acceptable to them or not, because this was considered by the League Committee as a foregone conclusion. The apparent intent in this line of inquiry was to enable the Committee at a later stage to fix the suitable distance of these zones.

Scientific reports on oil travel, however, proved conflicting and became a dilemma for British officials who were responsible for the policy recommendations on the width of the protective zones. Egypt, Sweden, the United States and Britain reported evidence or observations on the drift of oil at sea. Oily slicks were observed to have drifted at least 15 kilometres, according to Swedish experts, and at most 500 miles, according to tests by a British expert at Egypt's Alexandria University. The United States, as to be expected, forwarded Dr. Stroop's second experiments in 1927, showing how fuel oil spills disappear at sea. These American tests, however, were challenged by at least two independent British scientific experiments – one by C.H. Roberts of the Ministry of Agriculture and Fisheries and another by the Admiralty.[143]

Another problem arose from the summary of replies to the question of the compulsory fitting of separators on new or existing ships. An even number of countries favoured and opposed the modification of existing ships. But significantly, 18 countries, as opposed to 10, favoured the modification of new vessels in future, including such major maritime states as Italy, Japan and Norway. Both Japanese and British shipowners added that oily water separators were being increasingly fitted on their oil-fueled ships. Poland and Czechoslovakia, which did not reply to the other questions, also made the same point.

Here was an opportunity, therefore, to introduce the question of fitting separators onboard new vessels. Unfortunately, the League Committee concluded that the requirement could not be made because most major shipping nations (meaning Britain and the United States) would resist the requirement. Grimshaw explained that the replies to those favouring separators were padded with the support of countries with little or no merchant marine.

The wider provision of port reception facilities was also scuttled by the League experts, since, apart from Britain and the United States, few major international ports had such facilities. It was considered futile even to recommend the subject for it 'could not be put into practice in the immediate future.'[145] Nonetheless, Grimshaw privately told the Foreign Office that port facilities would facilitate the success of the zone system, and he eventually urged a recommendation on their provision as part of the League draft convention.[146]

British officials continued to maintain the fiction before the public that they had an open mind about the control of oil pollution. They submitted a judiciously-balanced summary of the views of shipowners and environmentalists to the League questionnaire, and their delegate to the 1935 meeting of the League Assembly pledged that Britain 'was prepared to accept any and every international project calculated to remedy the evil' of oil pollution.[147] So effective was the smokescreen as to confuse even the Admiralty at one stage.[148] In confidential exchanges, however, British officials were already preparing a draft convention based principally on the zone system.

The 1935 League Draft Convention

In September 1935 the League Council passed a resolution paving the way for the preparation of a draft convention and an international conference on oil pollution.[149] Grimshaw's Committee of Experts returned to Geneva and met from 21 to 25 October 1935 to prepare the draft treaty based on replies to the League questionnaire and subsequent discussions within the Committee.

Grimshaw presented the Committee with a second British draft text, mainly the work of the FO's Assistant Legal Adviser, W. Eric Beckett, and this draft was altered by the League Committee in only a few parts.[150] The 1935 draft League convention on oil pollution was stylistically and legally superior to the 1926 draft Washington convention, having been the result of years of refinement by legal minds in the British government. The main system of control for pollution was a system of zones in the high seas wherein oil or oily mixtures could not be discharged by ships belonging to the states which adhere to the treaty. Each contracting state was duly authorised to designate areas off their costs, the maximum width being 50 nautical miles in normal circumstances and 150 nautical miles for special coastal configurations.[151] A clearer method of fixing zones between state-parties was also incorporated in the draft treaty.

During the drafting of the League treaty, and subsequent negotiations for its approval, the thorniest legal question became the method of jurisdiction over the zones and over the ships of state-parties. After all, of what use were these pieces of paper if they could not be enforced?

Three types of legal jurisdiction were considered, namely, modified dual or shared jurisdiction, coastal state jurisdiction, or flag state authority over their vessels. Each system presented a host of problems which have always plagued the acceptability, and ultimately the effectiveness, of oil pollution agreements.

Had the views of British officials prevailed, the draft League convention would have introduced a system of concurrent or dual jurisdiction within territorial waters, such as was contained in Article VI (2) of the draft text given by Grimshaw to the second meeting of the Committee of Experts.[152] The reason for this novel sharing of sovereignty was, ironically enough, not so much to protect the coastal state from pollution as it was to ensure protection for ships of the flag state from arbitrary rules or wide territorial limits imposed by some states. British officials entertained visions of ever-increasing claims by coastal states for powers over foreign vessels, and they had to protect their extensive merchant marine from these hostile moves. In their view, there was a serious possibility that special zones for oil pollution throughout the world may turn opinion more in the direction of wider limits of territorial waters.

A shared system of jurisdiction within territorial waters assumed great importance for British officials in two ways. It would be a defence against the possibility of British vessels being caught polluting in territorial waters wider than the traditional three-mile limit. Secondly, it was also an assurance that Britain would not have to forego its right to punish offenders of its own domestic legislation by the act of ratifying the treaty. Thus, by appearing to strengthen the basis of enforcing the draft League convention, British officials were also gambling that they could extend protection for their vessels even within the territorial waters of other states.

The League Committee of Experts overwhelmingly rejected the British proposal and excised all references to shared jurisdiction in the draft League convention, mainly on the argument that sovereignty was indivisible. Many of the experts also expressed their anxieties about their ships being placed under double jeopardy, with more than one arrest and prosecution for one offence.

Secondly, the French Government submitted a proposal for serious consideration of exclusive coastal state jurisdiction in the prohibited zone.[153] According to a French memorandum submitted to the Committee of Experts, it was important to national sovereignty and security that supervision of the zone be 'exclusively' entrusted to the authority of the coastal state. To French officials, coastal state jurisdiction would deter foreign vessels from discharging oily waters off their coasts, which at the time were seriously affected by oil pollution. Secondly, the French argued that since there was no agreed definition of what constituted territorial waters anyway, they ought to legitimise the functional authority of a coastal state to protect itself within the prohibited zones.

Grimshaw was momentarily attracted to the practical value of the French suggestion for coastal state jurisdiction, but this was precisely the type of movement which the FO Treaty Department were trying to squash. Fortunately for the British, despite a serious pollution problem, French officials withdrew their own suggestion.

Thus, it was agreed once again that jurisdiction over the oil pollution zones would belong exclusively to the flag state, or the state having authority over the vessel.

Several legal safeguards were built into the draft treaty in order to anticipate its enforcement. State-parties would be required to impose 'adequate penalties' for violations of the zone system and to investigate all reports received from other parties as regards offences by their vessels. French officials insisted on the provision for imposing penalties in order to strengthen their government's proposed new legislation. Secondly, state-parties would 'watch over' and duly 'note offences' of ships within the prohibited zones. However, the right of coastal state invigilation over zones worried the

Admiralty and French officials, since the invigilation might be used as an excuse to hamper the movement of shipping, or indeed facilitate a military operation.[154] Thus, the Foreign Office promised to minimise or excise this provision at the international conference to finalise the treaty.

One of the more lasting practical suggestions for enforcement in the draft League treaty was the forerunner of the oil record book. Its inclusion in the proposed text was due to Dagfinn Paust, head of the Norwegian tanker owners association, whose presence at the second session of the Committee of Experts appeared controversial at first.[155] On Paust's suggestion, shipmasters would be required to enter into the ship's log all incidents involving the discharge of oil or oily mixtures in port or at sea. Paust drew the Committee's attention to the fact that ships carry a log book into which masters record all pertinent data of the ship's operation. Future oil pollution conventions would incorporate this provision.

The draft League convention contained other clauses referring to the central agency (to be established under the aegis of the League); the exchange of information on new control methods (physical or chemical treatment of oily wastes); the status of small vessels (to be decided by the international conference); naval vessels (exempted from the treaty); and a system of conditional ratification (whereby a state-party would accept the treaty only upon similar acceptance by some other state or states). The ratification of eight states would bring the treaty into operation.

Despite its legal veneer, the draft League convention really weakened the possibility of controlling oil pollution, firstly, by allowing maritime interests a wide margin to evade the terms of the treaty, and secondly by relinquishing governments of a previously-anticipated commitment to provide more meaningful answers to the problem. Its weaknesses flowed from many inherent provisions of the proposed treaty itself.

The zones would be determined in such a way as 'to take into consideration that vessels desiring to discharge oil should not be obliged to deviate too much from their normal route before reaching their port of destination.[156] Moreover, masters were given an exceedingly lenient excuse to plead that a discharge within the prohibited zones was caused unvoidably by an emergency, perhaps due to stress of weather or some other unspecified reason. Whilst the record of oily discharges in the ship's log was intended to assist in the detection of offences, this was really of small value, since masters who did not exercise care could hardly have been expected to give evidence against themselves.

Another inherent weakness in the draft League treaty was that it closed the door to the possiblity of encouraging ships to retain oily wastes onboard, even for new ships which might have been built according to pollution control specifications. As we have seen, the 1926 draft Washington convention had promised governments to ease the burden of and to encourage shipowners to equip their ships for the separation and retention of oily wastes onboard. But the 1935 draft League treaty shunted these provisions and turned them into 'recommendations' in the draft Final Act for the proposed League conference oil pollution, rather than incorporate them in the draft treaty which would have bound the contracting states.

The reasons given for this unfortunate decision were complex, technical, and self-defeating, but they were advanced by both the British officials who prepared the draft text and next by the Committee of Experts of the League who approved the final draft treaty.

The 1926 draft convention provided incentives on tonnage measurement reduction and the removal of any other disability for ships equipped for retention of oily wastes. But these incentives were deleted from the body of the draft League treaty by British officials who did not wish to implement them. At first, the Board and the Foreign Office did not agree on this score, because the latter saw no harm in deleting the incentives provision whilst the former saw no harm in retaining them.[157] The Board were advised that even for large ships, exemption for tonnage due only to the retention of oily wastes would not be large in comparison to the other deductions to which vessels were entitled. Furthermore, the reclaimed oil had commercial value.[158] The Foreign Office, however, concluded that it should be left to the discretion of the responsible authorities whether the question of incentives had any practical significance.

Since they had now sidetracked the issue of shipboard separation and retention of oily wastes, the policymakers had to placate environmental critics with a complement to the zone system. To make the draft League convention more palatable to domestic critics, and indeed to ensure compliance with a system of prohibited zones, British officials and League experts toyed with the idea of a compulsory requirement on the provision of port facilities for oily wastes from ships.

Port reception facilities for oily wastes had been discussed as the appropriate cure for oil pollution from ships since the start of the problem in the twenties. Ships complying with a system of zones, or even those equipped with separators, needed port reception facilities to handle the wastes which could not otherwise be re-cycled onboard or discharged at sea. It was as elementary as it seemed. But the most troublesome technical discussion at the Committee of Experts, as with previous and subsequent international considerations of oil pollution control, was the extent to which the various countries were prepared to provide these port reception facilities.

British, French and Italian members of the League Committee insisted that port facilities be made compulsory, whilst the majority, led by the Americans, resisted the compulsion. In sum, the main objections to a treaty requirement on port facilities were as follows: some ports were autonomous from the national authorities; port authorities in principle wanted the shipping and oil interests to solve the problem of pollution; only very few countries could afford the costs of these facilities; and, finally, this requirement was likely to lead to delays or even refusals to ratify the proposed treaty.

In order to accommodate both points of view, and with an eye to the public at home, Grimshaw presented a motion to insert a strong recommendation in the draft League Final

Act on the provision of port facilities, with the possibility of reviewing the situation after the treaty had come into operation. The Committee agreed to this compromise -- port reception facilities would not be required but only recommended, and it would be left open to reconsider this measure if experience showed that the system of zones was not working.

When the Committee of Experts had finished their work, the draft League convention and draft Final Act were circulated to governments on 27 November 1935, with the request that replies be returned to the League by 1 April 1936.[159] In effect, the League of Nation's work on oil pollution terminated soon after. Despite the overwhelmingly favourable replies to an oil pollution agreement, the draft League treaty was never signed, the proposed diplomatic conference never summoned, and the diplomatic negotiations drifted aimlessly.

Voluntary Action by Maritime Interests

During the twenties and thirties, voluntary measures taken by port authorities and oil and shipping interests accounted for a reduction in pollution. These voluntary measures were taken either in anticipation of similar action being required of them through national or international legislation, or because some economic benefit was gained from pollution control, or because these interests wished to maintain good relations with the public. For whatever reasons, these measures obviously reduced pollution, and they were commendable steps in the right direction. However, they induced a degree of complacency on policymakers for further legal measures.

A significant improvement in the situation in ports from the mid-1920s onwards was noticeable, due, as we have seen, to national legislations setting prohibited zones and also to the voluntary provision and use of port reception facilities in some major ports. The experience with coal pollution and its control by means of receptable barges for coal ashes became the precedent whereby large ports provided separator barges for receiving oily water wastes from oil-fueled ships or tankers. However, port authorities generally refused to be under any legal compulsion to provide reception facilities for several reasons: in principle, they wanted to throw the burden of solving the oily mess onto the oil and shipping interests; the expected volumes of oily wastes lay beyond the capacity of existing port reception facilities; and finally, they correctly believed that the zone system would drive the pollution away from ports onto other areas.

As the 1935 League survey showed, seven out of thirty-four countries responding to the questionnaire had ports providing reception facilities for oily wastes from ships. Several countries, both with or without port facilities, stated that they had requests pending for more such facilities. In the United States, the American Petroleum Institute had initiated intensive evaluation of the adequacy of oil terminal reception facilities and tanker practices to reduce the amount of oily discharges from American tankers as far back as the mid-1920s.[160] In Britain, a number of large ports had installed reception facilities dating to the early 1920s. By 1923, the Port of London Authority reported that they had two separator barges, worth £4,250 each, which had serviced eight ships. Altogether 75 tons of oil were reclaimed at a cost to the shipowners of about £505.[161] However, some port authorities with reception facilities reported that, in many cases, ships preferred to dispose of their oily wastes at sea rather than pay for the use of these facilities. Moreover, the British Chamber of Shipping noted that the campaign for ships to have compulsory separators seemed to have been abetted by the makers of separators for oily wastes, who naturally preferred the commercial possibility of providing for thousands of ships rather than for a few hundred ports.[162]

Had the technical controversy on the provision of port facilities been resolved, either by national legislation or international agreement, it was not inconceivable that the prevention of oil pollution might have been more successful from the start. But since that was not to be the thrust of control measures, the supply and the use of port reception facilities remained a poor cousin to the method of discharging oily wastes at sea.

Shipping interests also contributed to the reduction of pollution in several ways: by the voluntary adoption of a zone system on the high seas, by the installation of separators onboard ships, and by better ship-building methods.

After the 1926 Washington conference, Hipwood met British shipowners on 6 August 1926 to discuss the recommendations of the conference. He personally urged them to practice the zone system on the high seas, irrespective of whether other countries were prepared to do the same.[163] British shipowners soon got in touch with the International Shipping Conference which sent circulars to all their members in other countries requesting their voluntary observance of a 50-mile zone from any coast. Seven national shipping bodies from Britain, Belgium, Japan, the Netherlands, Norway, Sweden and the United States returned a favourable answer and issued the necessary instructions to their shipmasters.[164] Thus, shipowners in these seven countries requested their shipmasters to refrain from washing out oily wastes within 50 miles off the coasts of any country.

In March 1936, the United States Government requested the extension of the zones to 100-miles for American waters.[165] The American request had been prompted mainly by a domestic problem – local furore over accidents arising from makeshift barges carrying bootlegged gasoline through inter-state waterways. Thirteen bills were introduced in Congress, leading to the passage of the 1936 Tank Vessel Act, which decreed stricter construction and operational standards for American tankers.[166] On the international front, the State Department tried to hasten the conclusion of negotiations for the draft League convention and enlisted the support of American shipowners in urging for a 100-mile zone among the participating members of the voluntary zone system. It is known that British shipowners acceded to the American request for a 100-mile American zone, and at least one British shipping company advised their ships to observe the same for areas off the British coasts.

Norwegian shipowners, who usually took their cue from British colleagues, however, were disconcerted by the British compliance with the American request. Their spokesman, Dagfinn Paust, wrote the American shipowners expressing his fear that, if the Americans insist on a 100-mile protective limit, other countries might do the same. The extensions, he reasoned, would close a good many narrow waters, such as the whole of the English Channel, part of the North Sea, the Mediterranean and the Red Sea. They might extend the widths of zones to the extent that the oil and shipping trade might be impeded.[167] Paust appealed to Grimshaw as well, and finally, the Norwegian shipowners decided not to comply with the American request.[168] The American zone was later reverted to 50 miles when it was found that the wide zone was unnecessary.

Oily water separators for oil-fueled ships became more widely used after the passage of various national laws and the 1926 Washington conference. By the time of the 1935 League questionnaire, Canadian and Japanese vessels had gradually fitted separators onboard ships. Italian and German shipowners supported the compulsory fitting of separators on oil-fueled ships to be constructed in future, because they realized the value of this technological innovation. A 1933 Board survey indicated that 24% of British vessels had been equipped with separators, and there were various types (gravity, centrifugal, filtration, etc.) in the market.[169] Shipowners who had not similarly equipped their ships with separators alleged, however, that these devices were installed for 'propaganda' purposes and insurance premiums, but they were not really being used.[170]

Thus, as times grew worse for the shipping industry, shipowners regarded the compulsory fitting of ships for pollution control as financial anathema, and they prevailed upon policymakers for caution in changes of policy. The estimated cost to the industry of fitting separators for all British ships was now estimated to be £895,000 (£570,000 for tankers and £325,000 for oil-fueled non-tankers).[171]

Tankers were claimed as a separate category of vessels which need not install separators, because a tanker 'in itself could become a giant separator.'[172] It was explained that tankers bringing in black oils and transporting similar cargoes to another British port refrained from cleaning their tanks, but instead mixed dregs of the old cargo with the new.[173]

Modern shipbuilding methods also accounted for a decrease in oil pollution. In the 1930s, the old 'rivet-caulk-and-add-a-bit' method of building ships was replaced by the all-welded frame, thus reducing much of the previous leaks and breaking up of older types of vessels. The coating of tank surfaces also reduced the amount of oil adhering to wet surfaces.

Chemical solvents were used in Admiralty vessels since 1929. These solvents broke up the oily emulsions and separated them before the oil tanks were cleaned in naval port facilities, and the patent was more widely distributed in the mid-1930s. Thus, as a general rule, chemical cleaning of oily tanks was done on British tankers during the 1940s to 1950s. But thereafter, refinery policy dictated less contamination of the oil cargo, and the method was abandoned by the oil companies themselves.

There continued to be wide differences of views on where policy should turn. Environmentalists rained their complaints on governments and urged more effective legal checks to the problem than the zone system. The oil and shipping industry pointed to their efforts as sufficient complements to the zone system and as proof that pollution was being dealt with. Meanwhile, governments generally preferred to let the economics of the problem, or the natural environment as it were, to take its course.[174] Also, government officials acquired problems more pressing than trying to keep the seas clean.

Failure of Diplomacy

The next few years were crucial to the fate of the 1935 draft League convention. By November 1935, the League Secretariat were hopeful that an international conference could be scheduled for 1936. However, internal League changes and an unstable international political situation caused fatal delays to the draft treaty.

In 1936, the League's technical agencies were re-organized, and the Secretariat excused itself from undertaking work on the oil pollution conference. This delay, on top of the League's difficulties in the political front, so angered Grimshaw that he wrote them back, 'we must apparently take it as settled that there can be no international conference before 1937, if the League is in existence then.'[175]

He was soon to regret that indelicate note when, after the League had cleared the path for the preparation of the conference in 1937, British officials now insisted on an indefinite postponement until they had received the firm support of Germany and Italy.[176] Frank Walters of the League Secretariat, referring to *The Times* leader of 7 April 1937 blaming the League for its 'tardiness', wrote to his British friends in the Foreign Office in London to stir things up.[177] Ralph Stevenson of the FO promised immediate action, and the previously-neglected despatches to Germany and Italy were sent off.

British officials, who had drafted the treaty in such a way that it could come into operation only after all the major maritime powers had ratified it, seemed hoisted on their own petard. International response seemed favourable, according to the official inquiries of the League. Of the major maritime powers whose ratifications Britain considered essential, all but Germany and Italy had not replied. Belgium, Denmark, France, Japan, the Netherlands, Norway, Sweden, and the United States reported their willingness to attend a League conference on the basis of the draft convention. The other countries declining to participate were not considered significant.[178]

In 1937, the Japanese government, which had favoured and helped to draft the treaty, was asked to approach their Axis counterparts.[179] Although they were still 'profoundly interested' in the matter, Japanese officials noted that they were now critical of some (unspecified) parts of the draft League treaty, and they declined to approach the German and Italian governments. British officials did not take this to be discouraging, for they hoped that the Japanese would not refuse the oil pollution agreement, especially if Germany and Italy could be persuaded to come to the League conference.

Germany's attitude was difficult to ascertain, because it had kept silent on the 1935 League inquiry and to the 1937 British diplomatic note. Changes of personnel since the Nazi takeover in 1933 led to one opinion that the subject was not fully understood by the new German functionaries. Germany had opposed the 1926 draft Washington treaty, but its current policy seemed more open-minded. German shipowners expressed a willingness to talk things over and even accept a limited undertaking. At the 1933 International Shipping Conference meeting, they told other shipowners that they would accept a compulsory obligation to install separators for oily bilge washings.[180] In 1936, British shipowners consulted with their German counterparts, and Grimshaw was informed that the latter's opposition to an oil pollution convention had been 'withdrawn.'[181]

Although Germany was no longer a League member, it remained interested in technical co-operation in general, for example the April 1937 International Sugar Conference convened by the League in London. Both the French and the American governments had received explicit replies that Germany would attend a League conference and consider a draft convention on oil pollution control. It was also made clear that despite Germany's withdrawal from the League, the German government would not obstruct the framing of an oil pollution treaty under the League.[182]

In April 1937, the British Ambassador in Rome, Sir Eric Drummond, was requested to consult with Italian authorities about the draft League treaty and proposed conference, but he cabled back that in his opinion the Italians would not favour the idea at the time. Unless his colleagues considered it important to secure an early reply, Drummond advised that a delay would enhance the chance for a favourable answer. Foreign Office officials did not press the matter.[183] However, U.S. Ambassador William Phillips had no reticence about interesting the Italians on the subject. Upon being told of the joint British, French and U.S. demarche to countries like Italy, Phillips obtained a personal interview in

June 1937 with Count Ciano, the Italian Foreign Minister, specifically to discuss Italy's attitude towards the draft League convention and proposed conference. Count Ciano expressed some ignorance of the matter, but he did not show any reluctance to co-operate within the League on this technical subject, and he promised to look into the matter.[184]

Evidently, British officials, who were expected to lead on this matter, thought they could not extract firm assurances from the Axis powers, which themselves were making extensive political claims in international affairs. Favourable signals coming from other sources were not counted on as reliable or definite assurances. In fact, British officials were secretly preparing to relinquish themselves of the responsibility or the blame in case of failure. The Board prepared a list of various excuses they could cite.[185] Publicly, they continued to maintain their desire to have a treaty on oil pollution, mainly in order to stave off the criticism of environmental groups.[186]

By 1938, however, British officials realised that the diplomatic impasse had to be broken, and a review of policy called for, since the prospect of a League conference seemed unlikely. They were faced with two options: to drop the question altogether or to continue the negotiations outside the League. The first seemed appealing to them, but under the circumstances, exceedingly embarassing, and, given the possibility of some success, it was untenable to quit trying. Too many important public figures had indicated their interest in the conclusion of an oil pollution treaty. An international climate favourable to the conclusion of the treaty was apparent to other countries. British officials had little choice but to forge ahead and re-open the negotiations on another front.

When French officials realised that the prospect of a conference under the auspices of the League was fading, they contrived in June 1938 to obtain identical bilateral agreements between interested maritime powers. The idea, though hardly new to British officials who had considered it briefly in the 1920s, was taken as an indication of French concern for their particularly severe coastal pollution.[187] The Foreign

Office put the question before the other ministries. The Admiralty wanted no more steps to be taken to secure any oil pollution agreement, bilateral or multilateral. The Board did not agree to the French suggestion but wanted to pursue a conference away from the League of Nations. Meanwhile, the FO thought of the clever idea of inveigling French officials into calling for the international conference themselves.[188] In refusing the unofficial French suggestion for identical bilateral agreements, British officials posed the open question as to whether 'some government' might be prepared to sponsor the conference away from Geneva. But French officials deftly parried away this invitation and added that they would be quite happy if such an arrangement could be made as soon as possible, doubtless meaning that they expected the British to assume this responsibility.[189]

Before it could be decided to take any action away from the League of Nations, British officials had to ask the League. This was done not so much to expedite matters but to find any possible excuse to delay. As one FO functionary wrote, after all, the League Secretariat might prove 'obstructive'.[190] On the contrary, League officials were delighted, and they even offered the technical services of the Secretariat if Britain wanted to call for a non-League conference in London. The idea for a conference outside the League, after all, had first been broached by Frank Walters in 1937, when it looked like a League-sponsored negotiation would present barriers to the oil pollution treaty.[191]

The Foreign Office hardly showed great enthusiasm to sponsor an oil pollution conference in London when its work on other matters seemed so much more important. Moreover, it was observed that public agitation in 1938 and 1939 seemed low and coastal pollution reportedly diminished.[192] They were helped in stalling by Douglas Howard of the Southern Department, which handled Italian affairs, to 'wait and watch' the actions of the Italian and German delegates at the forthcoming London conference on aircraft fuel tax exemption.[193]

The results of the aircraft fuel conference in March 1939 were satisfactory, with approval by the German and

Italian delegations. The Italian delegate did not sign only for lack of a full power. It was felt that similar arrangements could be made for the oil pollution conference. By March, the Board and the Foreign Office considered the possibility of the 'bother and expense' of an oil pollution conference in London.[194]

As the British appeared to be dragging their feet, officials from Norway, Ireland and the League Secretariat lobbied diligently behind the scenes to promote the conference. Baron Metternich, acting director of the League's transit agency and secretary-general of the aircraft fuel conference, told the British delegation that, if Britain did not want to sponsor the oil pollution conference, Norway and Ireland expressed a willingness to do so. The FO conveyed the message to the Board without comment. It was now up to the Board to decide whether to step down in favour of Norway or Ireland taking the lead, or indeed, if London should be the site of the oil pollution conference.[195]

In June 1939, while maintaining that the question was important, the Board felt that they had their plate full of other work and that the decision on the oil pollution conference should be deferred for another six months. The FO seemed greatly relieved to have this response, and the files on th subject were put away until January 1940.[196]

The Second World War erupted in Europe soon after this last departmental exchange, and government officials naturally acquired more urgent problems than oil pollution control. Strangely enough, for the period when Britain seemed mainly removed from the Nazi advances on the continent, British officials continued to pass regular inquiries among themselves on the subject of the proposed conference. The situation seems even more absurd due to the fact that, on 26 August 1939, the Cabinet Committee on Imperial Defence had transferred the responsibility for the loading and unloading of all British merchant ships from their owners to the Ministry of Shipping and had authorised the Admiralty to assume control of merchant shipping at sea. Finally, on 3 December 1942, R.C. Williams of the Board wrote a brief note which captured the spirit of diplomatic initiatives in oil pollution of this period: 'I take it we may defer action on this question from year to year'.[197]

Yet again attempts to secure international control of oil pollution had failed. To some, it had seemed an opportune time to secure the adoption of a treaty, even if that treaty offered little more than a palliative towards solving the problem. But the Second World War nullified the possibility of diplomatic negotiations, and even had the technical treaty been achieved before the war, the hostilities would have led to universal denunciation of the agreement on zones.

Despite evidently good intentions, the international efforts to control oil pollution in the 1920s and 1930s ended in a history of abortive conferences and abortive committees. For this record, responsibility falls equally on those who disliked the idea of interfering with commercial pursuits for the sake of environmental protection, and others who reneged on their commitment to secure an international agreement. It is not entirely fair to say, as it has been accepted to date, that the international conference to finalise the draft League convention never took place because Germany, Italy and Japan refused to attend the conference. Britain and the United States, far from fighting to the last ditch for their initiatives and convictions, instead, allowed the negotiations to come to an inclusive end and in fact contributed to the obstacles and delays. It does not seem unusual moreover, for people, when they turn into a cul-de-sac, to plead that in their belief it was the fault of others, never their own.

4

THE 1954 INTERNATIONAL CONVENTION

International action against oil pollution by ships was finally achieved by means of the 1954 International Convention for the Prevention of Pollution of the Sea by Oil, the product of an international conference in London sponsored by the British Government. Through a combination of fortuitous circumstances, the 1954 Convention was drafted, signed, and adopted as the world's first working treaty on oil pollution control.

Policy During and After the Second World War

Oil pollution control policy during the Second World War was an important function of strategy and defence. Wartime shipping had to observe strict methods of reducing discharges from ships and tankers, in order to avoid detection of oil slicks by enemy submarines at sea.

In Britain, chemicals were used to separate and recover dried oil from oil tank sludges during the war years, in effect putting into wider practice what the Admiralty had discovered for naval ships since 1929. For a decade after the war, chemical treatment of oily wastes was regarded as a promising means of reducing pollution; one such chemical, Fomescol, became widely marketed from the late 1950s. Refineries, however, found that chemical additives interfered with the refining process of oil and oil products, and they refused further chemical treatment of oily cargoes. Later apologists for these companies pointed out that the wider use of chemicals on oily wastes would have led eventually to secondary pollution of ports and waterways. By 1961, it was established that research in Western countries would take a non-chemical turn.

The Soviet Union continued to adapt and improve upon chemical treatment of oily wastes onboard ships as a routine procedure. With some irony, Western oil companies presently market oil-based chemical solvents and emulsifiers in a commercial venture to cleanse oiled beaches and oil spills at sea, after these had been polluted by oil released from ships.

In the United States, wartime expansion of the tanker fleet was accompanied by various official requirements leading to a considerable saving of oily wastes. Apart from good housekeeping measures and seamen training programmes, American tankers deliberately resorted to deeper loading of tankers, to as much as four to six inches above the Plimsoll line, which enabled tanker retention of oily dregs and their delivery to the next port of loading, to be mixed with the new cargo of oil. Such a practice was also noted for British coastal tankers in the 1930s, as we have seen, and these operations went smoothly, with little difficulty and no incidents. American shipping officials and maritime interests appreciated the commercial possibilities of deeper tanker draughts, especially after their practice during the war, and again during the 1956 Middle East crisis. The United States, however, did not press this suggestion at conferences on load-line regulations, due to what they claimed as Cold War politics.[198]

British Initiatives

After the Second World War, British officials in the Board (now working under the wing of the Ministry of Transport) reviewed their policy on oil pollution on an annual, semi-annual, and finally, quarterly basis until 1948. Although they agreed that an oil pollution convention was desirable, they had quite a lot of other more pressing matters.

After the Geneva maritime conference in 1948, however, the establishment of a United Nations specialised agency on maritime affairs seemed imminent. The convention of the Inter-Governmental Maritime Consultative Organization (IMCO), was ready for signature and states could join the new maritime agency. Although IMCO did not in fact come into operation until 1958, British officials had to decide whether the new UN body should be used as the forum for securing an oil pollution convention in much the same way as they had gone through the League of Nations before the war. The troubled start of the U.N. maritime agency, however, gave pause to such plans. Even had IMCO started earlier, British officials did not contemplate immediately filing the problem of oil pollution before the new organisation. As Percy Faulkner, head of the Mercantile Marine Department advised: 'it would be better to wait and see how IMCO got along before placing the subject of oil pollution before it.'[199]

Meanwhile, British environmentalists launched a new campaign for international action on oil pollution. The Minister of Transport was asked in a Commons debate on 31 October 1949 to place the oil pollution problem before IMCO as soon as it had started work, to which the former could only reply that 'when the time came' the British Government would do so. The Earl of Ilchester, with some poignancy, wrote in *The Times* on 27 June 1949, 'Some progress was made towards this problem before the war. Can no further action be taken by the Government?'

From 1946 to 1949 complaints reaching the Board about pollution had escalated and seemed increasingly serious. For a while, it was maintained that no action would be taken until and unless shipowners themselves submitted a case for international action and environmentalists increased their agitation. These two conditions were met by early 1952.

In January 1952, the International Chamber of Shipping (ICS), formerly the International Shipping Conference, acknowledged that oil pollution was serious enough to merit a codification of rules. This was followed by a statement of the president of the British Chamber of Shipping in June 1952, admitting that the problem could be dealt with more effectively through an international treaty.

In March 1952, British environmental groups campaigning against the oil problem formed a co-ordinating committee (predecessor of the Advisory Committee on Pollution of the Sea or ACOPS), with James Callaghan, then an opposition member of Parliament and representing concerned interests in Wales, as its first president, and Miss Phyllis Barclay-Smith as its first secretary. Callaghan promised that their main objectives would be to campaign for an international convention to outlaw oil pollution and to urge more effective legislation in Parliament.

ACOPS sponsored the first international conference by a non-governmental body in London on October 1953, and invited representatives from governments, oil and shipping interests, port authorities, and environmental groups. At this unofficial conference the Transport Minister, Alan Lennox-Boyd, announced that Britain was 'speedily' consulting with other governments on the possibility of an international oil pollution treaty.[200]

However, the timing and the forum for such an international negotiation for an oil pollution treaty yet troubled British officials. In particular, they wished to stave off a reconsideration of the problem by the United Nations and IMCO.

Immediately after the war, the Transport and Communications Commission (TCC) of the U.N. Economic and Social Council (ECOSOC) had picked up the thread of the League of Nation's work on oil pollution control. It was also anticipated that IMCO, when it started operation, would directly handle the matter. When IMCO did not function immediately, the U.N. Transport Commission recommended that urgent and immediate consideration be given to oil pollution control nonetheless. The U.N. Secretary-General was asked to solicit the views of governments and render an official report, 'Pollution of Sea Water by Oil,' which was issued in December 1949. The report stated that there was a case for urgent international control of oil pollution, since 'conditions may on the whole be as bad or possibly even worse than they were a few years ago.'[201]

Member-governments were again asked in September 1950 to ascertain whether they were interested in taking action pending the entry into force of the IMCO convention. Replies to this inquiry showed that whilst the majority of governments experienced oil pollution, they differed on the next course of action.[202] The U.N. Transport Commission recommended that governments form national committees to study oil pollution control policy, and these national reports would be sent to IMCO. But if IMCO had not started work by 1952, then the Transport Commission would create its own group of experts to recommend further action.[203] By 1952, the Transport Commission was set to do precisely that – authorise a committee of international exports to draft an oil pollution treaty. But this move was indefinitely postponed after consultations between the U.N. Secretary-General and the British Government in January 1954.

Britain decided to take the initiative outside the aegis of the United Nations at this time, mainly in order to ensure the harmony of the oil pollution treaty with its maritime interests. By upholding the pre-eminence of maritime interests before environmental protection, British officials were only following a tradition set in the past whenever this problem was considered by governments. The strategy of earlier negotiations before the war showed that the lead country (in this case Britain) would take the strongest possible position at the conference, mainly in order to please a domestic audience, but, anticipating resistance from other major maritime powers, it would prepare a number of fallback positions more acceptable to its own domestic commercial interests. In order to do this, the problem of oil pollution had to be studied in light of contemporary developments in the oil and shipping industry.

The Faulkner Report

Significant developments in the maritime transport of oil had affected the problem of oil pollution to the extent that even some maritime interests appreciated the need for government intervention on the international level. These relevant patterns in the oil and shipping trade caused attitudinal changes towards environmental protection, but they also indicated the extent of the overall stakes held by the industry in resisting strong control measures.

The character and magnitude of oil movements at sea had altered substantially. By 1953 over 250 million tons of oil were transported by sea annually, as against 90 million tons annually in 1938. Half of the oil shipped (137 million tons) was now crude petroleum. Crude oils have the tendency to show more obvious soilage of beaches and surface waters than fuel or lighter oils. When crude washings were discharged from tankers, it appeared that the pollution had worsened in many parts. Thus, tanker discharges became the major object of concern, when before they were regarded as only a secondary source of pollution, after dirty bilge slops or oily ballast from non-tankers.

The growth in transporting crude petroleum was due to the policy of major European states which had encouraged refining of oil for home markets, rather than import already refined oil products from foreign sources. Oil had also become established as the major source of fuel for ships and industry, replacing coal. Almost 90% of ships were using oil fuel.

Whilst before the war the normal size of tankers ranged from the 5,000-10,000 ton capacity, in the 1950s oil demand brought orders for the 18,000-28,000 tonners, and the latter size was then considered as a 'supertanker' of the day.

The Faulkner Committee was appointed in 1952 to objectively examine the nature of the oil pollution problem and to recommend control measures. The ad hoc inter-ministerial group was called 'The Committee on the Prevention of Pollution of the Sea by Oil,' and placed under the direction of Percy Faulkner, chief of the Ministry of Transport's Marine

Division. The Report of the Committee, popularly known as the 'Faulkner Report', was published in July 1953 and attracted considerable acclaim, with some justification. It also became the keystone of British policy for the 1954 international conference.[204]

The Faulkner Report avoided the shortcomings of its predecessors, the 1926 American committee report and the 1935 League Committee findings, by addressing itself to highly technical but necessary explanations of the procedures whereby different types of ships causing oil pollution could refrain from dumping oily discharges at sea. Each method of control was extensively assessed, and its applications considered for particular classes of vessels in different circumstances of trade. Instead of pressing for a panacea in the form of a universally-applicable procedure or technology, the Faulkner Report reviewed selected methods of pollution control which had been tried and found to be technically and practically feasible by British operators.

The recommendations covered both short-term domestic and long-term international action. Immediate domestic action was sought by means of specific amendments to the 1922 British Act. The 1922 Act was to be revised, not only to increase fines and provide for stricter enforcement of its new rules, but also to enable shipowners to practice modern anti-pollution procedures within British waters.

With regard to their recommendations on high seas pollution, the Committee correctly brought past and present proposals for control into perspective, thus:

> Broadly, there are only two ways of keeping oil out of the sea; namely discharge of all oily residues ashore, or separation and consumption of the recovered oil in the ship. Each presents its own special problems.

These 'special problems' involved such considerations as the lack of adequate port facilities for oily wastes in many parts of the world, especially in the Middle East, now a major oil-exporting region; the (then) widely-held assumption that tanker dregs could not be re-cycled or mixed with cargoes of a similar type (though this was recommended and later became the basis of the Load on Top System in the 1960s); and the lack of special equipment in ships to process re-usable oil fuel.

Nevertheless, the Report examined certain problems and procedures for different classes of vessels which, if adopted, would prevent oil pollution, provided certain conditions were met. Tankers and non-tankers (dry cargo ships, liners, tramp steamers, etc.) could practice shipboard retention and separation of oily wastes due only to cleaning or ballasting by means of certain procedures and the installation of special equipment. If port facilities were available, these ships need not present any appreciable pollution problem.

Tankers

Tankers were identified as the major source of oil pollution. Both tanker procedures leading to the overboard discharge of oily wastes (cleaning and deballasting) were said to be capable of minimisation by the use of a 'slop tank' method during the voyage:

> When carrying out the tank cleaning process it is possible without causing pollution to discharge overboard a large proportion of the washing if they are first pumped into a "slop tank" and allowed to settle to effect separation of the oil and water by gravity. One or more of the cargo tanks can be used for this purpose. A similar process can be carried out with the ballast water in the dirty tanks. This process has been well-known to tanker operators for a number of years and it is in fact carried out in some cases where it is desired to avoid going outside the fifty-mile limit to discharge washings and the ship is proceeding to a port equipped with facilities for receiving this oily residue.

A 'negligible amount of oil' amounting to 50 parts or less of oil in 1,000,000 parts of the mixture' (50 ppm) would be pumped out during the last stages, as the oil and water interface in the oily water mixture is reached. But the discoloration of the water during the last stages of pumping out at sea would be evident to the crew, and the pumping can then be stopped, thus retaining most of the oily dregs on-board. Under favourable circumstances (e.g. calm seas, care on the part of the crew, etc.), the Report noted that upwards of 80% of tanker washings and ballast water could be discharged without causing pollution.

Non-Tankers

Non-tankers (dry cargo ships, passenger liners, tramp steamers, etc.) gave rise to pollution only when they used their fuel tanks for ballast purposes. But it was recognised that dual-purpose tanks used for fuel and ballast, and double bottom hulls used to mix fuel oil and ballast water, were increasingly becoming obsolete. The Committee noted that the increasing use of modern shipbuilding methods and of oily water separators reduced pollution from these ships.

Types of Oil

The Report limited its recommendations only to ships involved in the carriage of 'persistent oils', which were defined as crude, residual fuel oils, lubricating oils, tar oils, creosote and similar substances, rather than those with 'non-persistent oils' such as motor spirit, kerosene, gas oil, animal and vegetable oils.

Zones

As an interim measure and an immediate palliative to the problem, the Faulkner Committee recommended that full trial should be given to the practice of a zone system in the high seas wherein various classes of vessels would not dump their oily wastes. After examining past reports on the distance of oil drift at sea and the behaviour of different types of oil at sea, it was concluded that the prohibited zones must be as wide as possible, but no specific distance was mentioned. The Committee's own investigations showed that crude and fuel oil which have thinned out to an almost invisible film were yet capable of building up and causing oily deposits on beaches. Moreover, members of the Committee did not require much convincing of the extent of oil pollution, since, as one of them told the author, 'some of us had had first hand experience of the state of the beaches when we spent holidays in places like Cornwall! '

Yet, the Committee concluded, as experts had done in the past, that a zone system's successful operation was contingent on ships being able to deposit their oily wastes to port eventually. According to their Report:

> Although we recommend the adoption of a zone within which persistent oils should not be discharged, we recognise that full observance of the prohibition will not be possible unless arrangements are made to enable ships to contain oily residues which cannot be used on board, and for the reception of such oily residues in . . . ports.

Port Facilities, Separators, and International Co-Operation

There were however two important qualifications, or a series of conditions which had to be met, before oil pollution control policy could be assured. Firstly, there was the question of the provision of port reception facilities for oily wastes and the equipment of oily water separators in non-tankers. Allied with this problem was the question of the disposal of waste oil from ships by port and refinery operators. Secondly, the co-operation of oil and maritime interests and governments was essential in any policy affecting the carriage and transport of oil by sea.

The Faulkner Report correctly pointed out that their recommendations would reduce the problem of oily discharges into the sea into one of providing port reception facilities for oily residues from ships. Port reception facilities were regarded, therefore, as the sine qua non of a successful oil pollution control policy for ships. But the provision of port facilities brought up two problems of longstanding controversy: Who would provide the facilities? And how would the oily wastes be disposed of?

It was noted that tankers could more easily find acceptance of their oily wastes in the terminals and refineries which they frequented, to the extent that these were owned by the oil companies which also owned the tankers themselves. But there were independent operators, whose tankers were chartered occasionally by oil companies or other agencies, who stood to lose payload capacity if they reserved space for slop oil in their tankers. Since these charterers could little anticipate the certainty of refinery acceptance of their oily wastes, or indeed the compatibility of the slops with the next type or grade of oil cargo, the easiest method for these operators to take was yet the discharge of oily wastes at sea. In 1952, the Faulkner Report noted that 58% of takers bringing in oil into Britain were registered in foreign countries. The Committee recommended, therefore, that oil companies chartering these tankers should endeavour to apply the same code of conduct on independent operators as they would for their own company ships.

Tanker terminal facilities in Britain and the United States were generally regarded as adequate, but if wider usage of these facilities was to be recommended, especially in loading ports, then facilities had to be improved in the Middle East and other main oil-exporting regions. Five British oil firms affirmed the Faulkner Committee's determination to require oily waste facilities at the loading terminals under their control. Foreign terminals in the Middle East were said to be controlled by multinational companies and regarded as 'a separate matter'. British oil companies pledged that they would use 'their best endeavours with the companies with which they were associated to ensure the provision of the necessary reception facilities' in the foreign terminals. However, as we shall see, such a promise proved to be exceedingly suspect.

Port facilities for non-tankers depended mostly on the resources and conditions of the port authorities managing the harbours for general shipping. Most European ports were controlled by official or semi-official bodies, and it was expected that firm government directives to require facilities for non-tankers within these ports would bring them into line, provided the costs were not prohibitive. Indeed, British port authorities were now prepared, under statutory provision, to accept oily wastes from non-tankers. But there remained the critical issue of independent ports, and the United States later vetoed the idea of facilities for these ports as an insuperable constitutional dilemma for them.

The compulsory fitting of separators for non-tankers was similarly controversial. The Report listed the well-known objections to such separators, including costs, excessive size, over-reliance on the human factor, the technical impossibility of separating oils with a specific gravity similar to water. Despite these objections, however, the Report somewhat optimistically asserted that 'the fitting of separators [on non-tankers was] the only practicable method of preventing pollution' from such ships. Such optimism could only have come from countries where most non-tankers had already installed oily water separators, e.g. Britain, Canada, etc., and had developed commercially-saleable technology for promotion to other countries. However, a demonstration of the technique, arranged for delegates to the 1954 conference, had extremely embarassing results later.

If these problems about the provision of port facilities and separators could be met, then there remained the one question to which the Faulkner Report (and existing technology) could not give a satisfactory answer: What would be done with the oily wastes? Without a full answer to this question, a shipping problem would only be translated into a land-based problem.

Fuel oil from ballast water might be re-cycled, but even bunker fuel tanks collected heavy sludges that would contaminate the oily dregs. Until technology could cope with the fuel oil sludges, the Faulkner Committee admitted that there was little anyone could do but to dump these 'substantial accumulation' of oil sludge at sea, preferably outside the prohibited zones.

Tanker residues posed even more disturbing problems, but the Report asserted that oil refineries often accepted tanker dregs:

> at least five refineries are already prepared to receive shipments of tank washings, oil and water mixtures and emulsions, and

> other waste oils, and some are doing so at present. In order to provide an indication of the commercial possibilities of such arrangements, one of the oil companies arranged an experimental coast-wise shipment of 850 tons of tanks washings (which had been largely dehydrated) from a repair port to a refinery. After treatment of the washings at the refinery the oil company found it possible to pay the suppliers a substantial sum.

This was all very well as an 'experimental' arrangement, but existing practice was otherwise more cautious. In the first place, such a policy of refinery acceptance of tanker dregs would have been made farcical by an existing customs duty amounting to £10,000 in the form of an Indemnity Bond to guarantee that the refuse oil would not be re-sold. Secondly, since it cost refineries 2½ pence per ton to process waste oil, their common practice was to dump the oil into pits (for example in the Essex marshes) where it was said fires often resulted from spontaneous combustion.[205]

Tanker dregs could also have been mixed with the next cargo, if the new load was compatible. But such a policy would only be resorted to much later after other options had been tried and failed.

The Faulkner Report envisaged that the total prohibition of discharges from ships was possible, if not immediately then at least after a period of adjustment. But it was only possible to attain the desired objective with the full co-operation of oil and maritime interests and governments, to whom the proposed regulations were addressed.

Preparations for the 1954 International Conference

The theme of total prohibition of oil pollution at sea was publicly played in Britain, but for effect it needed more than a one-man band. Invitations were soon extended to some forty countries whose oil and shipping interests were involved, or who had taken part in the 1948 Safety of Life at Sea conference in London.

But developments in the United States cast an ominous shadow over the conference, so much so that one former delegate to the 1954 meeting recalled that the Americans were 'the most obstructive' participants.

American officials gave numerous reasons for their recalcitrance. They claimed that they did not have enough time to consult with their maritime interests. It is interesting to note that the U.S. also refused to participate in the 1953 United Nations group of experts, whose formation had been forestalled by the British Government's initiatives. Secondly, American officials were reluctant to impose new regulations on their domestic interests when the oil pollution problem in the U.S. seemed minimal.

The British had presented them with a fait accompli in recommending the compulsory installation of separators on non-tankers, which was sheer anathema to American shipowners. An American Shipping Board's project, after the 1926 Interdepartmental Committee Report, tried to construct a successful separator for oil-fueled ships in the U.S. But this was later abandoned by the Federal Maritime Commission (the Shipping Board's successor). Several separators installed in American non-tankers failed to function properly and had to be removed. Meanwhile the specific gravity of Bunker C fuel, which most American ships used, approximated that of water, and gravity separation for this oil was no longer sufficient to separate the oily mixtures from ballast water in fuel tanks.

Also, as far as American officials were concerned at the time, the solution to the international problem of oil pollution at sea largely depended on the implementation of national laws and the adoption of 'good housekeeping' measures by oil and shipping interests. Taking pride in the good effects achieved by just such an approach for their area, it was only natural that American officials would adopt a more cautious attitude towards oil transport regulation, when it seemed to them that domestic feelings were running high against such international interference.

Finally, American officials and maritime interests were convinced that the formation of national committees to study the problem and technological developments, rather than new regulations, held the promise of a solution to the problem of oil pollution at sea. According to one American expert who went to the 1954 conference, 'attempts to regulate without understanding were experienced in the U.S. during the Second World War. This so hampered the operations of our merchant marine and foreign vessels that the President had to void all of the earlier war regulations and issue ones which had been prepared by those who knew what they were doing.'[206]

American reluctance to negotiate an international treaty on oil pollution control made it appear that, whilst others thought such a treaty was helpful, those who had less to lose by reason of their smaller fleets were now preparing to inflict heavy sacrifices on the maritime powers who were better qualified to deal with the problem. The situation inspired one former American delegate to recall the words of Macaulay: 'For those behind cried "Forward", and those before cried "Back".'

Last minute attempts by British officials, including a personal visit by Faulkner and his aides to Washington, proved inconclusive. With the American position unclear, and the difficulties of convincing other important maritime powers about the seriousness of oil pollution, a leading British official summed up their expectations of the 1954 conference thus: 'We did not expect to get total prohibition then.'[207]

The 1954 London Conference on Oil Pollution

The International Conference on Pollution of the Sea, which was held in London from 26 April to 12 May 1954, marked the first diplomatic conference on the problem, with all delegations (except that of the United States) having full powers to frame and sign an international convention.[208] At the opening ceremonies, the British Transport Minister, Alan Lennox-Boyd, gave prominence to the fact that thirty-one countries or 95% of world shipping were represented at the conference.

The 1954 conference delegates elected their officers by acclamation: Sir Gilmour Jenkins, Permanent Secretary of the Ministry of Transport and head of the British delegation, became President; and Counsellor Gunnar M. Boos, chief o the Swedish Shipping Department and leader of the Swedisl delegation, became Vice-President. The British Governmen provided the secretariat and conference facilities, and, tc facilitate its work, the conference created various committee: and sub-committees of delegates.

In his opening statement as President, Sir Gilmour Jen kins described the main tasks of the conference as follows (1) the formulation of an international convention on oi pollution; (2) consideration of the types of oils which cause pollution; (3) practical restrictions on tankers and othe1 types of ships would solve the problem; and (4) definition oi the terms of enforcement for the agreement. A communication from the International Union of Official Travel Organisations, representing 21 countries, was also read to impress the delegates with the need for a rapid and effective end tc oil pollution.

Accreditation discussions threw the conference into the midst of Cold War politics, when the Soviet and Polish delegates asked for the representation of the People's Republic of China. Moreover, the Soviet delegate insisted on speaking in Russian, which problem was solved when the USSR delegation agreed to provide multiple translations of Russian statements. Western delegations however prevailed upon the rest to shelve the question of Chinese representation, under protest from the Socialist delegates.

The conference was initially slow to overcome the 'indifference' and 'ignorance' of many delegates. A British negotiator recalled that, 'The chief difficulty, at least in the early stages, was to convince the delegates that there was a serious problem.' Only eight countries admitted the seriousness of oil pollution in their area, as opposed to twelve which cited it as being an occasional trouble, negligible, or non-existent (see Fig. 2). It was the lasting impression of one American delegate that many of the participants were almost completely 'ignorant' of the subject which brought them together.[209]

It appears that a few countries sent delegates whose only credentials were their diplomatic status in their London missions. The famous Nicaraguan poet, Ruben Dario, for example, represented his country as its Ambassador and lone delegate.

British officials tried to overcome early indifference by flying some of the delegates to western beaches so that they might see for themselves the large chunks of oil which littered these areas. Apparently, this was effective, and some of them became quite 'enthusiastic' for effective measures. 'Ignorant' or not, the delegates had to listen to or study papers on the details of the problem and control measures by technical experts and leading delegates, and they had to record their preferences when the time for voting came.

A Convention or None At All

A few days after the conference had settled down, the General Committee (chaired by Faulkner) discussed the preparation of an international convention to embody the recommendations of the conference. It was then that the U.S. delegation tried to convince the rest that a treaty was premature. As an alternative to an immediate treaty, they proposed the creation of national committees to study the problem and its solution and the establishment of an interim international secretariat to review the possibility of a convention until IMCO had started to function. But the other delegates felt otherwise. D.C. Haselgrove (UK) summed up the general opinion thus: 'A convention is the customary and most satisfactory means of settling an international problem of this kind.' Captain Loennechen (Norway) held a convention essential to pre-empt the possibility of special measures by aggrieved coastal states. With the sole opposition of the U.S., the conference agreed to draft an international convention. In subsequent discussions, the American delegates continued to participate actively, but they abstained from voting when the proposals were formally introduced as draft articles.

To accommodate the Americans, it was agreed to schedule another conference in three years' time and to recommend the establishment of national committees on oil pollution. The United Nations Secretariat was asked to collect and distribute information on port facilities and research efforts in the countries concerned.

Oil as a Pollutant

The conference easily accepted that the oils to which future prohibitions would apply should be only those of the 'persistent' category (crude oil, fuel oil, heavy diesel oil, and lubricating oil).

There was considerable disagreement on the persistence of oil in the marine environment, with the Dutch, French, Belgian and American delegates trying their best to shorten the time during which oil remained capable of causing pollution. Discussions on the persistence of oil at sea were highly relevant to the kind of measures to be agreed upon, for if it could be found that oil discharged from ships posed a lasting threat to the oceans and littoral regions, then there was indisputable logic in the argument for taking the strongest possible measures. The British delegation led one opinion which called for total prohibition and pointed to the Faulkner Report's findings on samples of crude and fuel oils washed onshore. The Belgian and French delegates, also quoting from the same Report, rebutted the validity of these findings and advanced the view that not all the evidence had been submitted. However valuable these experiments may be, alleged the French delegate, they were only 'hypotheses on which it would be impracticable to set up a permanent system of costly remedies.' Mr. C. Moolenburgh (Netherlands) reminded the delegates of Dutch scientific tests to show the biodegradability of oil by marine organisms, of the 1927 tests carried out by Dr. D.V. Stroop for the U.S. Government, and of more recent Dutch studies on oil at sea, all of which convinced the Dutch Government that oil would not persist in the oceans. Mr. James E. Moss (USA) added that crude oil from oil fields contained living organisms which 'induced a gradual process of distillation within the oil, separating the heavier element from the lighter elements.'

From these discussions, at least one delegate who originally leaned towards total prohibition was swayed into accepting lesser measures. Fortunately, it was not the intention of those who believed that oil eventually disappeared due to natural processes, to torpedo the conference, as opponents of strict measures tried at the 1926 Washington conference. Instead, they determined to persuade the 1954 Washington conference that oil pumped at sea could be discharged when ships were a 'safe' distance from shore, and in this respect they were largely successful. These differences of views were reflected in Resolution 1 of the Final Act of the conference.

In any event, immediate trial for a system of zones seemed to reconcile the conflicting scientific data on the fate of oil at sea. For those who believed that oil lingered indefinitely in the marine environment, a zone system offered at the very least an immediate palliative. At worst, if the zones failed, then that would have been proof that the zone system offered no safeguards against pollution.

The conference only briefly touched on the subjects of accidents, war wrecks and natural seepages. For the most part, the 1954 conference followed the pattern of previous international discussions on oil pollution and only dealt with the question of operational pollution by ships.

They turned to the legal definition of what constituted harmful amounts of oil mixed with ballast and cleaning water. Once again, the legal definition of oil pollution was founded on the technical means then available to minimise, but not entirely prevent, pollution. The delegates accepted that shipping discharges of oily mixtures with more than 100 parts per million (100 ppm) of the oil in the mixture would be regarded as an indictable or administrative offence, if discharged in contravention of the other terms of the convention. This definition was an improvement by five times over of what the 1926 Washington conference had agreed were harmful amounts of oil in operational discharges by ships (500 ppm), but if fell short by half of what the Faulkner Committee had determined, through tests onboard the tanker *Narica,* that a tanker was capable of minimising oily effluents to 50 ppm.

How the tanker crew were to tell when they had reached the legal limit of pollution in the discharge was not en-entirely clear.

Types of Ships

The conference next turned to the question of the different regulations for various types of ships, with exemption being granted to war vessels and naval auxiliaries, whaling ships, Great Lakes coasters, and ships below 500 gross tons. Exceptions to prohibited ships under certain circumstances (e.g. emergency and lack of port facilities for a certain time period) were also granted.

Non-Tankers

For non-tankers (dry cargo, passenger liners, etc.) using oil as fuel and mixing ballast water in their fuel tanks, it was the considered opinion of the British delegation that the installation of oily-water separators was the best possible means of preventing oil pollution from these types of merchant ships. They were prepared to grant an exemption for ships using Bunker C fuel or for ships nearing obsolescence. But, in their view, 'the technical and economic objections to separators could now be disregarded . . . oily water separators are already fitted in many ships and various types have high standards of performance.'

Many other delegates, however, did not feel that a convincing case had been made for the compulsory fitting of separators, much less their use, on non-tankers. The classic litany of economic and technical defences against separators was recounted – times were hard for the shipping industry, expensive machinery would mean troubling an already highly competitive industry, a corresponding rise in price of oil products was untenable, and the installation of separators was no guarantee against malfunction or human error. Discussions at sub-committee stage came to no definite decision. The British continued to urge the requirement, but others (even those who favoured separators) proposed to let each country make its own decision on this issue.

Before they closed, a British expert assured the delegates that the feasibility and efficiency of separators 'would be convincingly demonstrated at a special arrangement of May 1st.' However, the demonstration turned into a debacle. As an American delegate recalled:[210]

> Those who knew of the history and limitations of oily water separators viewed with alarm the announcement by our British friends that they planned to demonstrate one of their separators on a British ship in the India Docks during the 1954 conference. It was a delightful affair – all but the separator. Nearly all of the delegates to the 1954 conference attended. The weather was fine, the luncheon and drinks was delicious, and the announcement raised all hopes. I stood at the starboard rail over the discharges from the separator [which was supposed to show a clean discharge] A few minutes later, oil started to flow into the river. I told Percy Faulkner. He followed me back to the rail. I felt very sorry for him. The demonstration was called off.

The Italian delegate tactfully noted for the record that 'the demonstration which took place last Saturday, was not definitely probative.'

Thereafter, the British delegation could not press for the compulsory fitting of separators on non-tankers. This was relegated into a resolution for the Final Act of the conference, but not as an Article in the convention.

As a viable alternative to separators for non-tankers, the British transferred their hopes to a decision by the conference on port reception facilities for oil wastes from these types of ships.[211]

Reception Facilities

The presentation on the indispensable value of port reception facilities was divided into two parts – for tankers and non-tankers. In the first place, it was recognised that oil companies, rather than governments, should assume the the burden of responsibility for tanker reception facilities in oil terminals. Tanker facilities were treated as 'a specialised matter' since tankers use a limited number of ports, and many are under the direct control of oil companies 'well qualified to deal with the problem.' British oil companies hastened to assure the delegates that, as far as they were concerned, they would do everything in their power to provide tanker reception facilities in their loading terminals 'regardless of cost.'

Thus, the provision of tanker reception facilities was not required of the very parties who confidently assured the conference that they were 'well qualified' to prevent tanker pollution. Tanker reception facilities became the subject of a resolution recommending them for the action of oil companies and ship repairers, but it was not made a binding obligation in the new convention. British environmentalists, in the parliamentary debates to enact the enabling legislation of 1955, tried but also failed to compel oil companies to meet their obligation. As Lord Hurcomb noted.[212]

> If it was so obviously in the interests of every oil company to provide these facilities for their tankers, why has it not been done in the past and why is it not being done all over thê world now?

Secondly, for non-tankers, governments were asked to undertake an obligation to provide port reception facilities for oily mixtures. In some ways the debates on non-tanker reception facilities reflected the same problems as for tanker facilities: administration, expense and technical difficulties. The Americans strongly resisted any such compulsion being made part of the new treaty, because they were not in a position to enforce it – U.S. ports were regulated by states rather than by federal law. None of the European countries shared the same constitutional dilemma.

From the conference proceedings, it seems that many states were swayed into voting in favour of non-tanker reception facilities under the assumption that the greater part of the latter's oily ballast water would have been separated out before it reached the port, for such was the conclusion of the sub-committee on port facilities. At the General Committee meeting of 11 May, however, the British delegate moved for a specific time limit on continued discharges by non-tankers at sea, provided these ships called at a port with reception facilities. When this was approved, Counsellor Boos proposed that non-tankers should be made to observe the same zones as tankers, if port facilities were available at their next destination, which also was approved.

These decisions altered the case for non-tanker reception facilities, since these ships could only begin deballasting operations when near or within the harbour itself, for the safety and seaworthiness of the vessel. If, as they were not being asked to do, they had to retain all their oily ballast water within tanker zones of fifty miles or more, and they had no separator onboard, then they had to retain all of their oily ballast water and bring them to port. Under the treaty, a state-party would have had to accept all of these oily ballast water. A number of coastal state delegations realized that they had committed themselves to something quite horrific, considering the potential volumes of oily water wastes they would have to accept. On the last day of the conference the Brazilian and Chilean delegates read prepared statements essentially indicating that they would not ratify the treaty if it contained the requirement on non-tanker reception facilities. An Australian proposal to enable port authorities to determine the percentage of water they would be prepared to accept as part of the oily wastes from non-tanker, which was supported by Canada and New Zealand, was initially defeated and reintroduced as a matter for the drafting committee. The drafting committee however reported out a text which was still unacceptable. This confusion came too late to prevent the inclusion of a binding obligation in the treaty for the provision of non-tanker facilities, and the issue was to haunt the enforcement of the 1954 convention. As we shall see, this compulsory provision on non-tanker facilities would be deleted when the treaty came up for review in 1962.

Tankers

As regards tankers, identified as the major source of oil pollution, the conference was apprised of several ways to control the pollution, as follows: (1) the use of chemicals; (2) cleaning at the port of discharge; (3) retaining oily residus and mixing these with the new cargo of oil; and (4) deballasting and cleaning as usual but retaining oily dregs in a 'slop tank' for discharge of the wastes at the oil loading terminal. The tanker sub-committee acknowledged that any one of these procedures depended on the goodwill of the oil industry especially, and that in many cases difficult and new problems would be encountered.

The use of chemical solvents or emulsifiers was disparaged. The second method, tanker cleaning at the discarge port, would have yielded a high degree of control, but the vast quantities of oily wastes to be handled meant unacceptable delays at port, an increase in tanker fleets, and vast expansion of storage facilities in areas where space was at a premium. The Faulkner Report summed up the technical and commercial problems in cleaning tankers at discharge ports, thus:

> Even if facilities to enable tankers to clean tanks and to discharge tank washings ashore were available or could be provided, the work might well add 3 to 4 days to the time at present spent in port. A tanker carrying crude oil to the U.K. makes some 10 voyages a year, and at least 30 extra days would be required for these 10 voyage. This would reduce the number of voyages which a tanker could make in a year and would necessitate an increase of some 10% in the tanker fleet of this country if the level of imports of crude were to be maintained. Moreover, in order to maintain the flow of oil imports, it is essential to make use of all available berth facilities, and as far as refineries are concerned berth capacity is one of the controlling factors in the amount of oil handled. This capacity would be reduced by over 60% if tankers remained at discharge berths in order to clean their tanks. Provision of special berths for tanks washings would seldom be practicable, and at many ports there is in any case no room to build additional berths. The use of anchorages when available at discharge ports for the purpose of cleaning tanks would involve serious congestion.

In such a way, British oil companies controlled the decisions leading to policy in the 1950s, as American oil companies had steered American policy away from this viable course in the 1920.

Next, if tankers refrained from cleaning their cargo tanks and mixed oily ballast residues with a new cargo oil, as suggested by C. Moolenburgh (tanker sub-committee chairman), then it was thought that this would lead to an unacceptable loss of cargo capacity if the next load of oil was incompatible with the previous dregs. The use of slop tanks within the tanker, complemented by reception facilities at the loading terminals, was said to be ineffective for journeys less than 5 days (due to the lack of time to allow for complete separation of the oil from the water), and the delays at port for transfer of the oily slops to shore (2 to 3 hours according to the British; 6 to 8 hours according to the Danish) left reason for caution.

Others felt that the fourth system (recommended by the British) would economically be self-defeating as well, in view of port charges, customs and canal toll duties imposed by some countries. But the most important disadvantage to the fourth system was the dearth of reception facilities at oil loading terminals, especially in the Middle East and South America. Not even the information on tests carried out by the British oil companies, which proved the viability of the use of 'slop tanks' complemented by reception facilities at discharge ports, could persuade those who were reluctant to act. The latter group was in any case composed of those who seemed convinced of the eventual disappeance of oil at sea and the attributed success of shipowners' practice of zones in the past, thus predisposing them to consider the ocean as a useful sink for shipping discharges.

The report of the smaller group of experts in the tanker sub-committee surprised the conference later.

Zones

Preliminary and sub-committee discussions on the zone system revealed significantly varied opinions on the width and enforcement of the zones. Some delegations pointed to the good results achieved by zones voluntarily observed by shipmasters off the U.S. coasts. Others countered that the beneficial effects of the American zones might be attributed to complementary measures by American interests and the general decrease of crude oil movement and gasoline from the U.S. to Europe. The French initially wanted coastal state jurisdiction of prohibited zones, and then later urged voluntary observance of zones. Some countries wanted the width of the zones to be determined by each individual state; others wanted a standard width. The British led one opinion calling for the widest possible zones, even for a 100-mile width as the normal distance. But this was defeated when the U.S. delegate summed up the general feeling that 'the mission of this sub-committee would be defeated if zones were made so wide that they amounted in effect to total prohibition.' The Soviet delegate repudiated the entire discussions as irrelevant and unnecessary and predicted that: 'the proposition would not have the desired effect, and a future conference would find itself faced with the same problems as the present committee was considering.' The Soviet opinion on oil pollution was entirely in keeping with their policy, for Socialist delegations to the 1958 United Nations Law of the Sea Conference in Geneva also held that the seas should not be used for dumping of any kind.[213]

Nevertheless, the zones sub-committee soldiered on, with those wanting stronger protection to maximise the widths of zones against others who tried to decrease the widths and application of zones.

Australia, the Adriatic, the Atlantic and the North Sea were designated as "special areas,' in which the 50-mile normal width (previously agreed upon) would be extended. The Australian zone was made as 150-miles from land, except in the north and west Australian coasts where a normal 50-mile width would be observed. The Adriatic, a closed sea, was strangely given a narrow belt of protection due to the Italian and Yugoslav delegates' joint wishes (to the exasperation of those who wanted zones as wide as possible) to leave a margin for their ships to continue cleaning their tanks at sea. On the last day of the conference, the Adriatic zone was declared to be 30-miles for tankers and 20-miles for non-tankers, these distances to be extended to the normal 50-mile widths within three years after the operation of the convention unless both Italy and Yugoslavia postponed such as extension.

The British and the French differed considerably on the fixing of an Atlantic zone for Western European countries, with the French being particularly adamant on a wide margin to be allowed for passenger liners and dry cargo ships on trans-Atlantic passage to discharge oily ballast. As a compromise, the Swedish delegate reduced the Atlantic zone coasts to a distance of 30° West longitude, instead of 40° W longitude (as the British had wanted), which decreased the area by about 300 miles but yet afforded wide belts of protection for most countries facing the Atlantic. Ironically enough, the French (who expressed fears lest the British later desire a wider Atlantic zone) subsequently proposed wider protection in the Atlantic at the 1962 conference.

The Swedish compromise was accepted at sub-committee stage. Britain asked for a re-consideration of the Atlantic zone at the General Committee, but their proposal for a 40° W longitude limit lost by one vote.

Wide belts of protection were given the North Sea countries by extending prohibited zones for this heavily-traversed shipping area to 100-miles from all coasts. Britain and West Germany jointly submitted a proposal for entire closure of the North Sea to oily discharges from ships, but this was defeated at sub-committee stage. Britain re-introduced the total closure of the North Sea to operational pollution at General Committee discussions two days later, and they secured a pyrrhic victory largely due to the support of countries which had no direct stake in the matter (see Fig. 2). Thus, on the following day the British delegate announced that they were voluntarily withdrawing their proposal for total closure of the North Sea to pollution. The majority of votes obtained in the conference, it was explained, would hardly sustain the obligation on states which had a direct stake in the matter and on whose support and compliance the success of total

Figure 2. State Policies at the 1954 Conference

			VARIOUS CONTROL MEASURES			
	Seriousness of the Problem	General Position on Total Prohibition	UK Proposal to Retain Tanker Slops until next Leading Port	Qualified Norwegian Proposal to Ban Tanker Discharges	Qualified Ban on Tanker Wastes Re-introduced by UK	Total Closure of North Sea to Oil Pollution
Country						
Australia		Yes	Yes	Yes	Yes	Yes
Belgium		No	No	No	No	No
Brazil		No	Yes		Yes	A
Canada	No	Yes	A	Yes	A	Yes
Chile		No	No		No	
Denmark	Yes	No	No	A	No	No
Finland		No	No		No	No
France		No	No	No	No	A
W. Germany	Yes	Yes	Yes	Yes	Yes	Yes
Greece	No	No	No	No	No	A
India		Yes	Yes	Yes	Yes	A
Ireland	Yes	Yes	Yes	Yes	Yes	Yes
Israel	Yes	Yes	A		A	Yes
Italy	No	No	No	No	No	A
Japan	No	No	No	A	No	No

	Mostly Not Serious	Rejection of Total Ban	Proposal Rejected	Proposal Carried	Proposal Rejected	Proposal Withdrawn
Liberia						
Mexico		A	No		No	A
Netherlands	Yes	Yes	A	A	A	Yes
New Zealand	No	Yes	Yes		Yes	Yes
Nicaragua						
Norway	No	No	No	Yes	No	No
Panama						
Portugal		No	Yes	Yes	Yes	
Poland	No	Yes	Yes	A	Yes	A
Spain		No	No	No	No	
Sweden	Yes	Yes	No	A	No	Yes
U.K.	Yes	Yes	Yes	Yes	Yes	Yes
U.S.A.	No	No	No	No	No	A
U.S.S.R.	No	Yes	Yes	A	Yes	A
Venezuela	No	A	A		A	
Yugoslavia	No	No	No	No		
*Tunisia	Yes				No	
Total Votes	8-12	12-14	10-14	8-7	10-14	9-5

Notes:
*Tunisia was an observer.

A = Abstained; Blank = No statement made.

closure, and that of the treaty itself, ultimately depended. Such tactic meant good international relations, but unfortunately it left a considerable part of the North Sea vulnerable to oil pollution, and the 1962 conference would later on be asked to close the 'hole' in the North Sea which the 1954 conference had allowed.[214]

A False Dawn for Total Prohibition Measures

Throughout the conference, there were some who, upholding the persistence of oil at sea and the availability of control measures, determined to persuade the rest of the delegates to commit themselves to total prohibition measures that would completely prevent operational pollution by oil ships. Even as the various committees were discussing proposals, the 'prohibitionists' prepared to marshall their arguments and influence the outcome of the technical working groups.

When the question was brought up before the General Committee, the problem of non-tanker pollution had largely been met through the compulsory provision of port reception facilities for these types of ships. Due to the width of the prohibited zones, non-tankers would have been compelled, if they wished to comply with the regulations, to bring their oily ballast water to port (unforseeable though the problem of enforcement was) or, even without a compulsory requirement to do so, to install separators onboard their ships.

The only outstanding problem to be settled was the more troublesome case of pollution from tankers, the major source of oil pollution. British, Swedish and Soviet delegates led an unlikely lobby favouring the strongest possible measures for tankers.

An initial advantage was gained when the tanker subcommittee reported that tanker retention of oily slops for discharge at the next oil loading terminal was a practical solution to adopt. At the General Committee, Faulkner drew the attention of delegates to the report and noted that by this practical procedure tankers could comply with complete prohibition of discharges at sea. The caveat was that oil companies had to provide reception facilities for tanker slop at the loading terminals.

The Greek delegate reminded the delegates, however, that the tanker sub-committee had been guided to a large extent by information provided by Britain, and the U.S. delegation had told the sub-committee that they would not be prepared to accept tanker retention measures. A Shell company executive tried to blunt the edge of this argument by saying that while the cost may be considerable, British oil companies had agreed to provide the facilities 'regardless of cost and they were prepared to honour this agreement.' Hence, the chairman (Faulkner) noted that it would be 'helpful to know' if oil companies in other countries were also prepared to do the same.

But when the crunch came, other oil company representatives, though admitting the technical feasibility of the procedure, refused to make any commitment at all. The French delegate moved for a quick ballot, but the delegates were given another day to think things over.

At heated discussions on the next day, the General Committee found itself torn between two opposing camps. On the one hand, delegates from Britain, West Germany, USSR and Poland argued in no uncertain terms for an early commitment to total prohibition. On the other hand, delegates from France, Italy, Belgium and Yugoslavia resisted anything other than the system of zones. Despite the best efforts of the champions of total prohibition measures and the logic of their position, they simply could not overcome an inherent bias on the part of their protagonists, a bias founded not only on the belief that there was a safe distance for ships to wash out at sea but on the reluctance to invest in facilities simply to conserve oily dregs and prevent pollution of the sea. According to the French delegate, Compte de Crouy-Chanel:

> While a general prohibition would be the ideal solution, it could not and should not take place at present. The evidence relating to the persistence of oil was not complete and furthermore possible prevention methods have not yet reached perfection. It was too

> early to draw definite conclusions or to impose a general prohibition particularly since such prohibition would involve extremely costly installation in ports and in ships. In considering expenditure of this kind, it was necessary to consider at what rate it was practicable to carry out such extensive capital investment; undue haste might have inflationary tendencies.

Lord Runciman, speaking as shipping director, tried to counter these arguments by saying that:

> The suggestion has been made that knowledge on this subject is not enough to justify a total prohibition; this argument applies even more to a zone system. It has also been argued that a total prohibition would involve too great an expenditure in providing the necessary arrangements in port and onboard ships but proposals that in conjunction with a zoning system certain facilities should be provided in ports did not seem to involve much less expenditure than would be needed in case of total prohibition. The establishment of zones would only reduce the problem of pollution by such small proportion that the necessary expenditure would not be worthwhile. Whatever the arrangements necessary for a total prohibition might cost now, it was certain that the eventual cost, if effective measures were not taken soon, would be even higher. Difficulties of enforcement did not seem to be an argument against the adoption of effective preventive measures . . . it would be much easier to detect violations of a total prohibition, where the absence of a propoer explanation as to how oil residues had been disposed of would provide evidence of an offence, than it would be to enforce a zoning system, where it might well be impossible to verify the exact place at which discharges had taken place . . .

The Belgians were disposed to take the easiest method. As their delegate pointed out, 'If discharges were allowed, facilities for tankers would not need to be provided so extensively.'

Thus, the issue of whether (and if so when) states would accept the principle of a total ban against oily discharges at sea was threatening to break up the conference. So irreconciliable were their views, and so adamantly did the protagonists hold their positions, that it was decided to elevate the matter before a Heads of Delegation meeting on the next day.

On 6th May, Counsellor Boos opened the discussion with the following statement:[215]

> I do not wish to return to my country without a formal convention providing for a genuine solution – complete prohibition. There is sufficient knowledge to enable the conference to draw up a reasonably practical convention. Public opinion would be on our side. A palliative would not be satisfactory.

The question which he posed was both complex and crucial. It opened huge tracts of doubts and questions which had seemingly been answered – by means of a compulsory provision of port reception facilities for non-tankers; by means of the oil companies' pledge to provide tanker reception facilities in loading terminals; by the zone system; and by a consensus for a trial period with zone until the treaty could be reviewed in light of technological developments or the environmental situation. To others, it was yet a question of principle and of timing – whether they would commit themselves now to a general prohibition, providing only for a transition period to enable the necessary tanker facilities and procedures to be organized.

Most delegates felt that, in any case, it would take several years for the treaty to operate. Canada proposed a specific date be given in order to encourage progress in the provision of facilities. Ireland wanted the date fixed at three years from the time of the conference. Australia saw no excuse for any delay, and neither did the Soviet and Polish delegates. Israel noted that the zone system was a 'poor substitute'.

But the opponents of total prohibition were powerful maritime nations. Admiral Shepheard (USA) repeated the American position: 'The United States cannot agree to total prohibition.' Belgium said the zone system was the only practical remedy at present. France, Italy, Norway and Yugoslavia also declared against total prohibition.

Leading members of the British delegation, sensing that the gains they had secured might now be lost by their grasping for more, accordingly moved to defuse the issue. Sir Gilmour Jenkins, as conference President, proposed the drafting of a 'strong resolution for the radical solution' to be placed in the Final Act of the conference and for the consideration of the problem by the next conference.[216] Effectively, this relegated the main objective of the conference into a hortatory set of words, rather than a potentially binding set of articles in an international treaty. The proposal was supported by several countries and accepted. Counsellor Boos maintained however that it was for the present conference to settle the matter.

Yet one more attempt, by means of a Norwegian proposal submitted to the zones sub-committee, was made to curb tanker discharges when port facilities were available at the next loading terminal. The Norwegian proposal, although decided by a slim majority at sub-committee stage, and re-introduced by the British delegation at the General Committee, was perceived however as 'an indirect means of securing total prohibition' and defeated accordingly. (See Fig. 2).

A 'First Practical Step'

The final plenary session was taken up by the presentation of the treaty and Final Act for the signature of the delegates and by the closing statements. The U.S. delegate repeated why they were reluctant to come but noted the success with which opposing views had been handled. He underlined the importance of the resolution on the creation of national committees and the study by the United Nations of the problem, and he promised the full co-operation of American interests in the observance of the zone system. The Soviet delegate called the convention 'the first practical step to protect the sea from oil pollution.' The French delegate thanked the British Government for their secretariat services and for the timely intervention of Sir Gilmour in preventing the conference from ending up in a deadlock.

Some delegates today recall that the 1954 London conference was a fruitful time and exceeded the bounds which most had hoped. Overall, the conference finally produced a convention which became the first working treaty on oil pollution.

The 1954 Convention and Problems of Control

The 1954 International Convention for the Prevention of Pollution of the Sea by Oil became the first working treaty among governments on oil pollution control. It took effect for contracting parties on 26 July 1958, some four years after it was reported out by the 1954 London conference. By means of the 1954 convention, environmental protection was officially recognised as an activity worthy of a binding commitment by the governments which ratified the treaty. Present-day analysts of oil pollution treaties generally deride the 1954 Convention, mainly because, in their judgement, it largely failed to control oil pollution from ships, the purpose for which it was intended. Such a view is abetted by official post mortem, since governments and maritime interests naturally are inclined to blame its limitations rather than their own inadequacies of policy and practice. It is well to remember that the 1954 Convention was the product of an international conference, a system which tends to reach for the lowest common denominator between divergent interests. Secondly, the problems of control derived not only from the limited terms of the treaty itself, but also from the conditions prevailing at the time, which determined the extent to which the obligations and recommendations produced by the conference could be put into practice. Finally, the state-parties to the treaty generally failed to implement and enforce the agreement.

To understand the extent to which the 1954 Convention was supposed to keep the seas clean, it is necessary ro recapitulate what it was not intended to do. In the first place, the 1954 Convention was not totally preventive of all types of operational pollution from ships. Its provisions were limited to some types of oil and to some control over operational pollution by certain classes of merchant vessels. Operational pollution from naval vessels, naval auxiliaries, ships less than

500 gross tons, whaling vessels, and Great Lakes coasters were outside the scope of the treaty.

After its limited scope, the 1954 Convention had problems inherent within the treaty itself, and prevailing conditions predisposed a tendency towards violation, rather than compliance. Broadly speaking, the 1954 Convention offered two means of controlling pollution from ships: (1) the compulsory provision of port reception facilities for non-tankers in the main ports of state-parties; and (2) the observance of a zone system for all relevant ships.

Port Facilities for Non-Tankers

Without an explicit requirement for non-tankers to retain their oily wastes from the use of fuel oil onboard, the measure seemed nonetheless inevitable due to the widths of the zones. After a grace period of three years from the operation of the treaty, all non-tankers were required to observe the same zones as tankers when they sailed to a port with reception facilities (Article III-2). To enable the retention onboard of oily wastes by non-tankers, state-parties were obliged to provide 'adequate' reception facilities in their 'main ports' three years after the treaty entered into force (Article VIII).

Opinion after the 1954 conference was mixed on the adequacy of the provision of reception facilities for non-tankers. Some thought that non-tanker facilities were not provided in adequate numbers even at main ports. Others, even those who paid lip service to the value of the binding obligation to provide these facilities, were convinced that the same Article prevented the early acceptance of the treaty by many countries. A 1961 IMCO survey showed that non-tanker facilities were not adequate. Many countries either did not reply to the survey or did not accept the 1954 Convention. Another IMCO circular addressed to member-stages requesting their acceptance of the 1954 Convention elicited disturbingly few replies.

Article VIII of the 1954 Convention anticipated a high degree of control for non-tanker pollution, mainly by enabling these ships to bring their oily wastes to port, even though the exact meaning of what constituted 'adequate' facilities and 'main ports' was left for each state-party to determine. As such, it was at the same time one of the strongest and most controversial parts of the 1954 Convention.

Because of the binding nature of Article VIII, and also because the provision of shore facilities is the sine qua non of any policy which seeks to prevent operational pollution by ships, the compulsory provision of port reception facilities for non-tankers was an integral part of the 1954 Convention. But it was precisely this article which prevented or delayed the acceptance by important maritime states of the treaty. The United States, in their delayed ratification of the 1954 Convention in 1961, only accepted the treaty subject to the reservation that it would not oblige their port authorities to provide such facilities.[217] The reservation was primarily intended to overcome the American constitutional dilemma whereby, under the principle of separation of powers, state governments rather than the federal government were charged with the regulation of ports and land installations. In any case, port facilities in the U.S. were widespread and generally regarded as sufficient. However, the American reluctance, delay, and ratification with an important qualification of the 1954 Convention had a demoralising multiplier effect. Some 43 out of the 55 ratifications of the 1954 Convention came after the U.S. acceptance, and about 15 of these came only immediately after the U.S. action. Liberia and Fiji made the same reservation to Article VIII of the 1954 Convention, in much the same words as the U.S. government's note to IMCO. Evidently, the U.S. position was crucial, and as we shall see, their ratification was made with the express intention of deleting the troublesome article at the 1962 conference, which had been scheduled by the time of U.S. ratification. The French government, in a note to IMCO on 8 March 1962, considered the U.S. reservation to Article VIII unacceptable.[218]

Interestingly enough, many countries which objected to port reception facilities due to the expense involved in such an obligation, nonetheless, complied with Article VIII and ratified the treaty.

As we have seen, the provision of reception facilities has been a longstanding controversy ever since the problem of controlling oil pollution from ships began in the 1920s. It raises fundamental questions about the sharing of obligations between coastal authorities and the maritime industry. Coastal (port) authorities have always taken the position that the problem of shipping pollution was a matter for the maritime industry alone, whilst the latter have argued for the necessity of an exchange of concessions.

Port Facilities for Tankers

The binding Articles on reception facilities excluded tanker facilities, primarily to accommodate the lobbies by the oil industry who owned most of these terminals. Thus, it was then accepted that tanker reception facilities would be a specialised matter for the exclusive concern of the oil industry, and it remained to be seen whether these interests would fall in line with the desired objective of preventing oil pollution. Resolution 4 of the 1954 conference 'urged' oil companies and shipyards to provide reception facilities for tanker wastes at oil loading terminals and repair ports.

British oil companies, which had given advice and pledged their commitment to provide reception facilities in terminals under their direct control, regardless of cost, had to take back their word and gave various excuses for their unseemly retreat. The provision of tanker facilities was not easily applicable to all oil companies. Although most tanker terminals are owned by oil companies, some, in the products trade, are owned by European governments and only leased to the companies. Problems of administration were evident not only for tanker facilities but also for all types of ports under local rather than national jurisdiction.

It seemed feasible, in principle, to clean tankers at oil loading terminals before they loaded the next cargo, since most loading ports already handle the salty oil-in-water emulsions from the oil fields which naturally mix with the crude from fields. But, it was said that to apply the same practice to tankers bringing in tons of oily water wastes required considerable expansion of existing facilities. Moreover, safety problems have to be considered for offshore terminals where submarine lines extended from the tanker to the loading terminal.

An even more immediate problem consisted of the division of multinational oil companies into different companies and into different departments within the same company. As James E. Moss, a former oil company executive described this accounting complexity:[219]

> It is usual to find that within the same company the group responsible for the operation of ships is not the same group responsible for the operation of the marine terminals. Indeed, in the case of a large company, four or even more groups, each having individual responsibility, might be involved in furnishing two services: sea transportation by tankers and shore accommodations for the oil wastes resulting from the operation of tankers. Thus, while there could be no question but that expense for the operation of the tankers would be for the account of the marine department there might be some question as to the account which should be charged with the expense for adequate waste oil facilities ashore.

In the light of later investigative reports of the multinational oil companies' tactical ploys, these administrative, economic and technical excuses would now seem suspect.[220] But at the time they were not challenged by governments. Of course, it was entirely possible for the oil companies to act independently of rival firms. But, as a spokesman for the British oil companies told the 1959 ACOPS conference in Copenhagen, the practice 'presents a problem which bristles with difficulties.'[221]

When the bill enabling the British government to ratify the 1954 convention came before the House of Lords, Lord Lucas of Chilworth urged the government to impose a statutory obligation on tanker reception facilities. If the oil companies had voluntarily assured the government about providing these facilities, he argued, 'then what harm is there in putting it into the Bill? ' Secondly, he noted that it was 'a flimsy proposition' to 'pass legislation on promises from prospective offending parties.' But the government spokesman, the Earl of Selkirk, replied that it was 'unnecessary to take these additional powers'. Any national measure taken in advance of internationally-agreed standards had to be weighed against similar concessions by other states for their ships and commercial interests. Lord Runciman's remarks illustrated this opinion well:[222]

> When we discussed this matter certain countries showed much less apprehension than our own about the quantity of oil that was going about the world and about the effect of oil. It would be unfortunate [to] to impose regulations which prohibited British ships from discharging oil in waters in which the ships of other flags were perfectly free to discharge it . . . it would not be fair, and . . . the cause which we all have at heart would not be helped, and might even be hampered, if this country were thought to be the general cleaner-up of other people's doorsteps.

The Zone System

The 1954 convention had specified another control system – the establishment of zones throughout the world. These zones consisted of belts of water off the coasts of all countries, being 50 nautical miles in normal width and varying in width for special areas (1954 convention Annex A). Certain exceptions were attached to the zone system. Dry-cargo, passenger liners and other non-tankers were allowed, if they found it necessary, to wash out 'as far as practicable from land' during a transition period, and, three years after the operation of the treaty, they were required to observe the same zones as tankers when sailing to a port without reception facilities (Article II-2). Ships' bilge slops were excluded for the first year after the entry into force of the treaty, and lubricating oil slops were entitled to an indefinite exemption (Article V). Sediments, or solidified dregs from tanker hulls and other heavy residues from oil fuel and lubricating oil in fuel tanks, were allowed to be discharged as far from land as practicable (Article IV-1-c). Emergencies could override any other considerations for observance of these zones (Article

IV-1-a and b).

Even had the zone system been successful in keeping the oil from coastal areas, it had the dual effect of authorising continued oil discharges at sea and of transferring the pollution from one area to another. An IMCO survey in 1962 showed that oil pollution had occurred in places previously unaffected by the problem. Oil company executives admitted that the zone system largely shifted the pollution from northwest Europe to the Mediterranean.

Moreover, it was widely suspected that the zones were more honoured in the breach than in the observance, especially by tankers, even convention fleets, washing under cover of night.

Two major difficulties in the zone system were identified: the difficulty, if not impossibility, of proving the offence, and the reluctance of flag states to proceed against their ships even when complaints were supported by evidence.

The main means of checking illegal discharges within zones were to be based on the oil record book notes by the shipmaster. In the oil record book, the master was asked to record all cleaning, deballasting and oil transfers, the date and the time, the place and the position of the ship, and 'approximate quantities of residues' (1954 convention Annex B and Article IX). Against assurances by a British official that the oil records were 'very detailed' and 'in a form which it is not easy to "cook," the French government later observed that the 1954 convention was impossible to enforce due to 'falsification of oil record books.'[223]

Flag State Jurisdiction

Convention ships were technically bound to observe the requirements but were subject to the final authority of their flag state. Flag state jurisdiction opened the door to a number of abuses and anomalies regarding violations of the zones. These problems, fueled by the fact that aggrieved coastal states had no powers of redress directly against foreign ships outside territorial waters, led to increasing disparagement of the system of flag state jurisdiction in maritime administration. Significantly, the 1950 International Law Commission meeting considered, but rejected, a proposal for a 200-mile coastal belt of pollution protection, mainly because of the anticipated international conference on oil pollution, which was expected to provide coastal states with just such protection.[224]

The problems which have usually been attributed to flag state jurisdiction of the terms of an oil pollution convention were clearly evident in the 1954 convention. Illegal acts of pollution were difficult to detect and to prove to the satisfaction of flag authorities. Uneven penalties were imposed by various flag states, despite the provision for uniform severity of penalties (Article VI), undoubtedly due to the reluctance of flag states to regard pollution in other places as being of equal seriousness as pollution within their immediate vicinity. Effective flag state jurisdiction called for a highly-developed relationship between the ship, its crew and operators and the national authority registering the vessel, which, in the case of 'flags-of-convenience' (FOC) vessels, was minimal when states imposed lax registration laws.[225]

Coastal state powers over convention ships were highly curtailed in the 1954 convention. Coastal states had only two functions as regards convention ships which they suspected of pollution within their prohibited zones. One was the right of inspection of the log or oil record book if and when the ship came to its port or to the port of another state-party, without however causing undue delay to the vessel (Article IX-2). Unlike the 1935 draft League convention, the 1954 treaty was silent on the right to invigilate zones.

Little wonder that the actual reported violations to the Convention within zones were few and far between. By the time of the 1961 IMCO survey, state-parties to the 1954 Convention reported that out of 92 violations in prohibited zones, none had been prosecuted successfully by the flag state. The record with regard to offences within territorial waters was similarly unsatisfactory: out of 577 reported cases, only 294 were successfully prosecuted, with largely puny fines.[226] Various governments submitted numerous technical and legal difficulties of control: e.g. obtaining definite proof that the oil discharged was of the persistent category (Denmark, the Netherlands, Sweden and Britain); or if the permissible level of pollution had been exceeded (Germany and Britain); or if coastal surveillance of the zones had been sufficient (Canada); or if log books had been falsified (France and Norway).

Secondly, coastal states could furnish proof of violations by vessels of the flag state 'wheresoever the alleged contravention may have taken place' (Article X-1). Whilst appearing broadly prescriptive of coastal states' rights to protest acts of pollution by flag state vessels, the relevant passage nevertheless, in many cases, restricted rather than enlarged the scope of powers of aggrieved coastal states. Indeed, despite an explanation by the British delegate to the 1954 conference that the relevant national law would apply to offences within territorial waters of state-parties, many governments continued to perceive questions on whether a party to the 1954 convention had foregone its right to punish violations even within its territorial waters. For example, U.S. ratification to the 1954 treaty was given with a reservation to specify the exclusive application of U.S. laws to offences within U.S. territorial waters.

If a flag state was a party to the 1954 convention and did not enforce penalties on the shipmaster or operator, the aggrieved coastal state might be regarded as having waived its right of punitive action for offenders within its own jurisdiction. Seen in another context, the aggrieved coastal state might have enjoyed greater protection within its territorial waters when it was not a party to the 1954 Convention but applied its own national law against the foreign ship, or if the coastal party had relevant legislation and the flag state was not a party to the convention.

This was indeed a very strange state of affairs, and it brings us to the question of whether flag states were now prepared to seriously accept their responsibilities for ships under the 1954 Convention's prohibitions against oil pollution. Flag state jurisdiction requires a close relationship between the port of registry and the vessel. But increasingly, many operators have chosen to register their vessels in countries with lax registration laws. These flags-of-convenience (FOC) vessels often cause concern about the enforcement of international maritime regulations, and the debate is yet open.

States with traditional control over their vessels did manage to prosecute against their ships for infractions in the territorial waters of another party when they were called upon to do so. But the puny fines imposed against shipmasters drew attention to the fact that even traditional maritime powers preferred gradual adjustment by their operators rather than strong punitive action against offenders.

It was expected, of course, that the operation of the 1954 Convention would serve as the start of an educational campaign for cleaner seas and good housekeeping measures onboard ships. To this end, the 1954 conference adopted a resolution calling for the preparation of manuals on oil pollution control. The 1954 Convention itself was later acknowledged as having provided an outstanding 'international code of conduct,' acting as a psychological deterrent for shipping operators to dump wastes at sea. But whilst the 1954 Convention may have instilled twinges of conscience against such acts, it certainly did not prevent oil pollution of the seas.

Law of the Sea Issues and Special Measures

Even if enforcement of the zones could have been adequate, the exact definition of these areas left much to be desired. Significantly, the 1954 Convention did not carry official charts for these prohibited zones. Moreover, it was framed when questions on maritime law regarding the extent of waters of national jurisdiction were highly unsettled.

At the final plenary session of the conference, for example, the Yugoslav delegate emphasised that the Adriatic zone must be measured from the base line of the inner line of its territorial sea. This formula was not followed, and Yugoslavia withheld ratification of the convention until after it had been amended in 1962, when a more acceptable formula for the measurement of the prohibited zones was agreed upon.

The assumption seems inescapable therefore that the limits of the prohibited zones were regarded as more or less extensive depending on the shipmaster's discretion, if he might have wished to follow the required procedure.

Considering the pre-occupations of traditional maritime powers like Britain over special measures, the 1954 Convention contained a specific Article limiting such measures only to ships of the state-party taking the special measure (Article XI). Thus, states wishing to specify measures of their own outside the scope of the 1954 Convention could do so with competence only for their own national fleet and within the limits of its jurisdiction.

Several parties to the 1954 Convention declared special measures exceeding or below that of the convention. Britain required British vessels to observe the Atlantic zone up to the distance of 40° west longitude. Some 200 vessels under Irish registry were allowed by Ireland to discharge oily ballast water from bunker fuel tanks within the prohibited zones. And the Canadian prohibited zone was extended for Canadian vessels to 100 nautical miles due to the declared grave ecological threat to the Newfoundland and Canadian Atlantic coasts from oil pollution.[227]

Conclusion on the 1954 Convention

Ironically, an international agreement to rid the world's oceans of oil pollution apparently contributed to the legitimisation of continued operational discharges by ships at sea. It only entered into force on 26 July 1958, some four years after it was drafted by the 1954 conference, due to what some argue may have been the conditional ratification procedure's delay (Article XV), which specified that not less than ten states, including five of countries each with not less than 500,000 gross tons of tanker tonnage, would bring the treaty into operation for the state-parties. By the time of the 1962 amending conference, only seventeen states had ratified the 1954 Convention, a few of which were at least as interested

5

THE 1962 AMENDED CONVENTION, LOAD ON TOP AND THE 1969 AMENDMENTS

The 1962 amended Convention provided for conditional total retention of oily wastes in all new large ships after June 1967. However, the oil and maritime industry found that they could achieve the semblance of reform short of what was expected of them. The Load on Top system was introduced on a concerted scale in the early 1960s, represented as the 'cure of operational pollution' by tankers, and substituted for the discharge standards of the 1962 amended Convention by means of the 1969 IMCO amendments.

Movement to Amend the 1954 Convention

It was taken for granted by participants to the 1954 conference that they would meet again in three years time to review their work. In view of the marginal reception of the treaty by states, continued or transferred pollution in many areas, and the real or perceived problems of control with the treaty, the amendment of the 1954 Convention became a necessity.

Early in 1957, however, the British government informed the other states that, since the 1954 Convention had not entered into force, it was premature to convene the next conference on that year. Only a year before, a United

Nations survey showed uneven concern and provision for reception facilities in many countries.[228]

On 26 July 1958, a year after the deposit of the French instrument of ratification, the 1954 Convention came into force for the ratifying states. Six months after, the IMCO Assembly, meeting for the first time in January 1959, accepted the bureau functions of the 1954 Convention from the British government. Counsellor Boos was appointed an IMCO consultant to prepare for the pollution conference which IMCO, this time, would sponsor. However, IMCO was more pre-occupied with work on the 1960 Safety of Life at Sea conference and convention, and oil pollution took second priority.

Official review of United States policy towards the 1954 Convention, which was sorely needed and wanted, started only after the 1954 treaty took effect and IMCO began its operations. The U.S. national committee on oil pollution control approved a draft report recommending that the U.S. accept the 1954 convention with certain reservations and understandings. Through their membership in IMCO and their acceptance of the 1954 convention, the U.S. evidently wished to assure themselves of a principal role in determining the shape of future control measures. As American officials told an unofficial gathering in 1959, 'We would like to propose amendments to the convention from the standpoint of being a member to it'.[229] American ratification was widely commended, and as expected attracted the adherence to the 1954 treaty of other major maritime powers, notably, Liberia, Greece and Panama. But, as we shall see later, American acceptance of the 1954 convention was not given without its price.

Meanwhile, world demand for oil during the economic boom of the 1960s nearly doubled, and world tanker and shipping fleets correspondingly increased. The closure of the Suez Canal in 1956 removed one of the constraints to the feasibility of building large vessels, and that year brought a flood of orders for 100,000 deadweight ton-tankers to haul crude around the Cape of Good Hope. In 1962 Japanese shipbuilders launched what was then considered to be an engineering marvel, a 130,000 dwt tanker. The de-colonisation process after 1960 led to the introduction of more new states in international policy-making, as well as the growth of new national fleets.

The movement for amending the 1954 Convention on oil pollution gained momentum and adherents. ACOPS and James Callaghan presented a seven-point plan calling for a worldwide ban on oil pollution. In a widely-publicised statement, Callaghan branded nations yet outside the 1954 Convention as 'too barbaric and uncivilised to bother'.[230]

A new conference fulfilled the need for fresh consideration to be given to the problem, whilst those who wanted to re-negotiate the 1954 treaty hoped to attract more support for their views. In 1961 IMCO issued invitations to more than 100 governments to attend a conference in London in the spring of 1962.

The 1962 IMCO Conference on Oil Pollution

Delegates from thirty-nine countries met again in London, under the aegis of IMCO, from 26 March to 13 April 1962 to improve the 1954 convention at the third International Conference on Prevention of Pollution of the Sea by Oil.[231] To give the decisions of the amending conference the status of amendments to the 1954 Convention, a parallel conference of contracting parties (a second conference in fact) took place and met on only two occasions, 4 and 11 April 1962. The mini-conference of contracting governments formalised the procedure for approving the decisions of the larger conference, and gave them the status of amendments to the 1954 convention. IMCO provided the secretariat for both assemblies, headed by Acting Secretary-General William Graham.

The British Minister of Transport, Ernest Marples, delivered the welcome address on 26 March. Mr. Marples noted that since the 1954 conference world movement of oil had doubled. He also drew attention to the tremendous expansion of world shipping tonnage (from 97 million gross tons in 1954 to 136 million gross tons in 1962) and world tanker fleets (from 25 million gross tons to 44 million gross tons). Finally, he attributed whatever success the conference would achieve to the 'will and determination' of the delegates.

As in the 1954 conference, Sir Gilmour Jenkins was unanimously elected President of the 1962 conference. The conference selected as vice-presidents the heads of the French, U.S. and Soviet delegations, as well as various committees and their chairmen.

Political Squabbles

Some eighteen states from Western Europe, North America, the Soviet Bloc and Australasia dominated the conference, whilst six states from Asia, four from Latin America and four from Africa hardly spoke at all. Only sixteen participants had ratified the 1954 convention, whilst the rest were expected to decide amendments to a treaty they had not accepted (which included the USSR and Yugoslavia, some of the more active delegates at the 1962 conference). For a technical conference, the 1962 assembly included a number of political skirmishes between the Soviet bloc and Western states and the newly independent states and traditional powers.

On the second day of the conference, the Soviet bloc delegates questioned the presence of delegates from Nationalist China, West Germany and South Korea, and protested the absence of representatives from mainland China, East Germany, North Korea and North Vietnam. Other political discussions involved the language issue, the adoption of an 'all states formula' for accepting the amended convention, the colonial article, the role of the International Court of Justice in dispute settlement, and the functions of IMCO.

The Soviet delegate insisted on an official translation of the amended treaty into Russian, which was not carried at the time. As a compromise, it was agreed that the Final Act of the conference would be translated into the four official languages of the United Nations.

The Soviet bloc delegates also wanted the acceptance of the amended treaty to be opened to all states, against a U.S.

proposal to limit potential convention parties to members of the U.N. and its specialised agencies. Although restrictive in character, the U.S. proposal to limit adherents to the amended treaty was carried by 15 votes to 8 against and 8 abstentions at the Legal Committee.

A joint Egyptian-Liberian proposal to revise the colonial article to take account of U.N. mandated territories was approved, also over the protest of the Soviet delegate.

The Soviet proposal on dispute settlement fell foul of the Western powers' regard for the International Court of Justice. In wanting to give state-parties a degree of flexibility in settling disputes over the treaty, the Soviet bloc delegates proposed a three-stage process of dispute settlement: negotiation, arbitration, and, only after failure of the latter, recourse to the Court in The Hague with the consent of the parties. The Soviet proposal was taken up repeatedly by the conference and defeated each time it was re-introduced.

Eventual Soviet ratification of the 1962 amended convention would contain reservations to both the colonial article and the sole recourse to the Court at The Hague for dispute settlement.

IMCO, which took over the functions of the central bureau in 1959, was also caught in political cross-fires as regards the expansion of the role of the organisation and other matters relating to the improvement of awareness of oil pollution and the amendment of the treaty in future. Alliances were drawn according to states' conception of the early role of IMCO, with the U.S. being particularly desirous of enlarging the uses of the organisation, and others, especially the USSR, trying their best to delimit its functions. The Acting Secretary-General found it necessary on several occasions to intervene and draw attention to the limited resources and the consultative nature of IMCO, then the youngest and smallest specialised U.N. agency.

A British proposal to oblige states to furnish IMCO with details of evidence against ships alleged to be polluting in contravention of the convention was opposed by the USSR and other states with large shipping commitments. The French delegate noted that the procedure might give 'an impression of suspicion' on the acts of other states. In view of these and other comments, the British withdrew their proposal.

Prompted by a 1954 resolution and a draft British resolution to more closely define the functions of IMCO, the U.S. advanced a proposal to turn IMCO into a central clearing house, responsible for the collection and distribution of various government reports on oil pollution. The U.S. proposal was found to be unacceptable due to its length and substance. Moreover, other states were simply not prepared then to create a permanent panel of oil pollution experts within IMCO. The Soviet delegate expressed some concern on the extent of the information required by the U.S. proposal. Finally, after much haggling over the extent of information to be given to IMCO and its function as a central agency, it was agreed that IMCO would receive any information supplied by governments but not necessarily disseminate it, complemented by a resolution on the provision of regular reports by governments to the organisation.

But the most important innovation for IMCO was the acceptance of a U.S. proposal to use the organisation as a means of amending the oil pollution treaty in future. When unanimous agreement by state-parties or negotiations through a diplomatic international conference was not possible, the U.S. delegation envisioned a procedure in future for amending the convention by means of recommendations put forward by the IMCO Maritime Safety Council and approved by the all-member IMCO Assembly. This was accepted by the 1962 conference, and such a mode of amendment to the oil pollution treaty was later to be used in 1969 and 1971.

Definition of Ships

It was decided that the amended treaty should contain a more accurate definition of both 'ship' and 'tanker'. A French proposal to define ships was found to be the most acceptable, since it applied even to barges, lighters and dracones pulled by tug boats, and it suggested that the responsibility for any pollution arising from the non-self-propelled ships would lie on the operator of the tug boat. A joint British-Norwegian definition of tanker was also accepted.

In dealing with the proposals to lower tonnage limits, the conference distinguished between tankers and non-tankers. Since it was generally acknowledged that tankers contributed the most pollution, the conference agreed to apply the convention for tankers of 150 gross tons, rather than 500 gross tons in the 1954 definition, while retaining the former level of 500 gross tons for non-tankers.

The old question of exemption for warships and naval auxiliaries resurfaced, with all delegates being agreed on the virtues of ending the blanket exception given to government military ships. Finally, whilst it was agreed to strengthen the obligations of governments over their naval vessels, a majority yet refused to place such ships in the same general category as commercial vessels.

The U.S. delegation observed that whaling vessels were permitted to discharge persistent oils and take other actions in contradiction of the convention, and they accordingly moved for the deletion of the former exemption granted to whaling ships. Norway, however, said that if this proposal were carried, they would be faced with the choice of having to leave the convention or wind up their whaling industry. As a compromise it was accepted that whaling ships would be included in the scope of controls when they were in transit to and from their whaling grounds, but not necessarily when they were on actual operations.

Kuwait alone suggested ending the exemption for ships traversing the Great Lakes waterways, though with what motives Kuwait felt called upon to safeguard the Great Lakes from pollution remained unexplained. Due to U.S. and Canadian insistence that the Great Lakes and the St. Lawrence Seaway were their special concern, the conference agreed to continue excluding vessels in these areas.

Definition of Oil Pollution

The 1962 conference discussed various proposals intended either to include more prohibited oils in the definition of pollution or to strengthen the means of enforcing requirements on shipmasters.

A working group on sediments and lubricating oil in ships' bilge slops favourably reported out a French proposal to include these slops in the scope of restricted substances. The group chairman noted that, whilst he was initially unconvinced, various samples were shown to prove that sediments tended to re-appear elsewhere from the place of discharge. The French also convinced the conference to include lubricating oil from bilges in the scope of the amended convention, due to studies on the highly toxic properties of lubricating oil.

The Yugoslav delegation moved for amendment of the definition of heavy diesel oil based on their viscosity rather than the standard method of distillation previously accepted. This had some justification inasmuch as it was not acted upon due to another proposal by the Dutch delegate for a certificate on the nature of the heavy diesel oil to be presented to bunkering stations. As a compromise, it was agreed to reconcile differences on these points in a conference resolution.

But the most revolutionary proposal to amend the definition of prohibited oils came from the British delegation's suggestion that the meaning of oil be taken to include 'oil of any description'. They were not trying to enlarge the scope of the convention, it was soon pointed out. There would still be a category of prohibited oils and oily mixtures. But the British proposal would have enabled any master caught discharging within zones to be taken to court and asked to prove that he did not discharge oil or oily mixtures in contravention of the terms of the convention, thus facilitating compliance with the amended treaty.

At the time, it was held that oils discharged from ships were almost impossible to detect, much less prove to the satisfaction of a foreign court that the permissible level of pollution had been exceeded. Attention was also drawn to the fact that many state-parties had reported the difficulty of proving that the oil discharged was of the prohibited category. The British delegate urged its approval, citing the deterrence it had proved for British ships since passage of the 1955 Oil in Navigable Waters Act by parliament.

However, none of the other countries were willing to reverse the burden of proof as a matter of legal principle and established habit. The Greek delegate noted that the British proposal would create more problems than it would solve, in case it was abused by coastal states. After maintaining the value of their proposal, the British delegation was forced to withdraw it.

Enforcement

The British also tabled a proposal for limited port state authority, which would have enabled another state-party to assist in the investigation of an offending vessel visiting the latter's port but not at the port of the aggrieved coastal state. There was some merit to this proposal in order to define more explicitly the procedure found in Article IX (2) of the 1954 Convention, which gave coastal state-parties the right to inspect oil record books. Despite the firm opposition of the Soviet delegate, the British proposal was passed by the Legal Committee.

Over the committee chairman's attempts to rule him out of order, the Soviet delegate apologised for re-opening consideration of the British proposal which they found to be entirely unacceptable. In short, it appeared that the Soviet delegation was prepared to walk out of the conference if the British proposal on port state investigation was carried.

The Soviet protest did not hinder the progress of the controversial proposal, at least not in the early stages. A revised British text was adopted by the Legal Committee and affirmed by the General Committee. But support for this proposal gradually eroded due to backstage discussions. The British delegation voluntarily withdrew their proposal in the afternoon of that same day's favourable discussion. The idea of port state powers was revived by the Canadians at the 1973 Conference, with some modifications.

It was inevitable that criticisms of the enforcement of the 1954 convention would imply a rebuke of exclusive flag state jurisdiction over vessels in prohibited zones. With the codification of international law of the sea by the 1958 and 1960 Geneva conferences, Belgium raised the possibility of enabling coastal authorities to board and inspect vessels of a ship belonging to another party if the latter were suspected of violating the prohibited zones. As the Belgian delegate noted, similar rights over ships on the high seas were allowed by states for pirate ships suspected of piracy, which was 'no more than an historical memory and constituted a much lesser danger than pollution of the seas.' Upon being apprised of the futility of their proposal being carried, however, the Belgian delegate withdrew the proposal even before it had been discussed by the Legal Committee.

Control measures could yet have been strengthened by means of ensuring that penalties for pollution acted as a deterrent. An American inquiry and reports to IMCO by state-parties for successful prosecutions of offending ships showed that in many cases the penalties imposed were not commensurate with the gravity with which the set of pollution was publicly acknowledged. Various proposals put forward to more accurately define the extent of flag state responsibility for their ships under the amended convention were discussed, rejected, or otherwise reduced to quite minimal proportions as they passed through the legal threshing floor of the conference. Finally, a French proposal (which was only a little more advanced than the original provision) on penalties was approved. This provided that the fines imposed should be 'adequate in amount and severity to reasonably discourage any repetition of such violation.' The drafting committee later altered this to mean that penalties should 'be adequate in severity' to discourage illegal discharges and false entries in the oil record books.

Before the French proposal on penalties was adopted, the Italian delegate underlined the basic premise on which they were working: 'that all infringements of the convention committed on the high seas should be punishable under the laws of the territory whose flag the ship in question was flying.' Any further clarification, alteration, or substitution of flag state jurisdiction was thought to involve the participants in 'too much detail' as to be unworthy of their attention.

As we have seen, the rights of coastal states over foreign vessels polluting near their shores would be resurrected time and again whenever the legal issues of enforcing marine pollution control measures were discussed.

Non-Tankers

Whatever the defects in the legal framework of the treaty might have been, there still beckoned the possibility of achieving the desired end to pollution by means of technological requirements and the enlargement of zones throughout the world. After the debacle at the 1954 conference, the British government evidently decided not to press other countries into accepting an obligation to install oily-water separators onboard non-tankers with fuel tanks used for ballast. Except for their attempt to maintain the utility of separators to prevent bilge leakages, which on the advice of the U.S. delegation was deleted, the British delegation did not submit a proposal for the compulsory installation and use of separators for non-tankers. Instead, they preferred to channel their hopes on separators through a resolution (which was approved) enabling IMCO to set suitable performance standards for oily water separators.

The discussions on anti-pollution measures for non-tankers involved three major issues: (1) the grace period allowed contracting parties; (2) provision of port reception facilities; and (3) the use of interchangeable tanks for carrying or storing fuel and ballast water.

Australia was particularly eager to extend the grace period for non-tankers to new contracting parties, allowing these ships to enjoy the privilege shared by previous state-parties. Others felt that the time had passed when countries could afford to wait and see the effects of the treaty, and hence once new parties accepted the amended treaty they should put its rules into immediate effect. However, the Australian proposal seemed practical as a means of attracting more states to ratify the amended treaty. A joint Australian-Norwegian recommendation was also passed to ensure that new parties who enjoyed the three years' exemption for non-tankers were encouraged to waive this exemption for themselves and ensure that their ships observed the relevant rules.

The compulsory provision of port reception facilities for non-tankers, being one of the strongest planks of the 1954 convention, became the subject of numerous proposals mainly intended to weaken it. Thus, Article VIII of the 1954 treaty was considerably weakned by a verbal subterfuge. Instead of state-parties being obliged 'to ensure the provision' of these port facilities, they would now be asked to 'take all appropriate steps to promote the provision of facilities'.

The American delegation scored a major success in persuading the 1962 conference to remove the compulsory obligation on governments to provide port facilities for non-tankers' oily wastes. In this respect, the British delegation floated an even more radical proposal – to delete the entire Article VIII altogether because it had prevented a number of countries from accepting the convention. The U.S., on the other hand, merely desired to ask all state-parties 'to take all appropriate steps' to have these facilities provided. Additionally, the state-parties would now be asked to include tanker loading terminals, repair ports, and a realistic assessment of which ports should have these facilities. But the new Article really relinquished them of the obligation to provide these facilities. Thus, it was made to appear that the U.S., by loosening one of the main planks in the 1954 convention, was actually making a concession in not asking for its deletion and in expanding the scope of reception facilities to include tankers terminals and repair yards as well.

A special working group on port facilities was asked to deal with the numerous proposals. The group reported favourably on the American proposal to encourage rather than oblige governments to provide reception facilities for non-tankers, tankers and ships under repair. Oil loading terminals for tankers and ship repair yards were included in the scope of the amended treaty, and ports were asked to provide suitable reception facilities 'according to the needs of ships using them,' rather than direct the measure only for the attention of main ports. Where facilities were considered inadequate, provision was made for parties to make their complaints through IMCO. The new article on port facilities for tankers and non-tankers alike was supplemented by a strong resolution urging their provision 'as a matter of urgency' and for periodic review of the situation by IMCO.

If port facilities were regarded as an invaluable tool against operational pollution by ships, the Greek delegation proposed an international fund to aid in the expenses involved in the construction of these facilities. The fund, which would have been established under IMCO's. aegis, would derive its income from a minimal levy on oil importers and distribute it to countries less able to bear the financial burden of providing adequate port reception facilities. Part of the income could also be used to subsidise the preparation of oil pollution manuals and the education of mariners. The Greek compensation fund seemed entirely worthwhile, at least to the French delegate who was also interested in advocating a related scheme to compensate ships which retained their oily residues onboard. The French envisaged the creation of a private international association with a dual purpose: to collect a uniform tax on all ships of state-members, and to distribute its funds as a bonus for ships which fulfilled the requirements of the treaty or as compensation for ships penalised by retaining oily residues as dead freight. The French scheme seemed even more attractive in view of the imposition by some countries of customs duties or canal tolls on oil slops or oily ballast water. As the French delegate said, 'It was not right that only the ships of the contracting countries should bear the expense of the measures for the prevention of oil pollution.' However, as the Greek delegate acknowledged, it was somewhat late for the conference to accept such radical measures, and both the French and Greek proposals were not supported.

The use of fuel tanks for water ballast in non-tankers was a practice which shipowners were asked to depart from 'if possible'. In addition to being a means to prevent pollution, the U.S. noted that the above procedure was also a safety requirement in the 1960 International Safety of Life at Sea Convention, and as such it was recommended to the delegates. But there was yet another way in which non-tankers contributed to pollution – by mingling water ballast with fuel oil stored in double bottoms as the fuel was spent. The Greek and Norwegian delegates requested the conference to prevent such a practice by means of a resolution. Unfortunately, the Greek delegation did not formally introduce their proposal, and the Norwegian proposal was combined with a similar recommendation to ban the use of cofferdams (at the aft and fore ends of cargo spaces in tankers) for settling oily ballast water, which was in fact a common legal practice. In the face of strong resistance from the other delegations, the Norwegian delegate withdrew their proposed resolution.

Tankers

Whatever the losses might have been in the control requirements for non-tanker pollution, the 1962 conference yet made headway in the measures to be required of tankers, the main source of pollution. The conference made substantial advances in preventing pollution from tankers in two ways: (1) creating new prohibited zones and extending previous ones; and (2) requiring total tanker retention of oily residues under certain conditions

Prohibited Zones

By consensus, the participants acknowledged that the 1954 zones were not meeting the problem of oil pollution from tankers. Studies by French and West German scientists validated the claim that oil at sea persisted and drifted ashore, causing a potential or actual threat to marine resources, wildlife and beaches.[232] In contrast to the 1954 conference, when many participants complacently accepted that oil discharged far enough from shore would eventually disappear and not cause trouble, the 1962 conference showed new respect for the scientific or actual claims that oil from ships at sea posed a serious problem to coastal regions and the marine environment. Only the U.S. delegation came forward with scientific evidence on the biodegradation of oil at sea, and, despite the considerable effect of this scientific research on future policy, it exerted little or no influence on the motivations and the decisions of the 1962 conference.[233]

By calling attention to the 1958 and 1960 Geneva conventions on the law of the sea, the German delegate secured the approval of a proposal to define prohibited zones more accurately as starting from the base line from which the territorial sea was measured, hence vindicating the Yugoslav position in 1954. The Canadian delegate expressed concern at the introduction of baselines, and as a compromise proposed that the measurement of prohibited zones should begin from where the territorial sea was measured, whether or not the state in question accepted the relevant articles of the 1958 Geneva Convention on the Territorial Sea and Contiguous Zone. But this was not carried. The Australian delegate disliked the use of the term 'as far from land as practicable' in designating certain zones. Tankers within the Great Barrier Reef might suppose that they were entitled to clean and deballast their tanks therein. An Australian proposal to insert verbal changes to cover this possibility (by adding 'in any case not less than 25 miles from land') was not supported and no action was taken.

With reasonable certainty that they had given a more accurate definition to the fixing of zones, the delegates now considered the feasibility of officially marking the zones in mariners' charts for inclusion in the amended convention. Great care was taken at the conference to mark the zones' correct longitude and latitude positions in order to avoid later difficulties and disputes.

Despite the strong desire to have wide zones, a Canadian motion to extend the standard width from 50 to 100 nautical miles became an early casualty during the discussions. Interestingly enough, the argument against adoption of a 100-mile zone as the normal width of protection was not that these zones were inherently unwise, but that they might in some cases be difficult to apply, and it was better to attract new parties to the amended convention by means of narrower zones. As the British delegate pointed out:

> it is dangerous to agree at this stage to general extensions of zones to 100 miles, because these zones would be applied to countries which had accepted the convention but would not necessarily be practical from the point of view of facilities. Secondly, there is a danger that other countries, having secured the benefit of the 100-mile zone, might not be persuaded to ratify the convention.

The Canadian proposal for general extension of all zones throughout the world to 100 miles, which was carried by a slim majority (2 votes) at its first consideration, was later nullified on a procedural misunderstanding, and then overwhelmingly defeated on a second decisive ballot.

Similarly, a Soviet proposal for a 100-mile zone for tankers and a 50-mile zone for non-tankers was withdrawn. The French and Madagascan delegations wanted the conference to consider allowing extensions of normal zones to 150-200 miles as the limit for general zones. But this was supported by only the British and Canadian delegations at the committee on ships. The width of 100 miles was retained as the possible limit to the extensions of normal zones.

Special Areas

Sympathetic responses were given to all requests for the enlargement of existing zones in certain areas and the creation of new special areas where total prohibition would apply. Two conditions were applied to the fixing of larger zones or special areas: that evidence be given of ecological damage or other circumstances requiring greater protection, and that the ratifications of coastal states would be a precondition for the setting up of special areas or extended zones.

A special working group was created to reconcile the proposals on extending zones. At one point, the Portuguese delegate complained that the committee leaned more favourably towards requests for larger zones from countries which had already ratified the 1954 Convention.

The French delegation wanted to attract attention to the amended convention by a most radical step – to take away the zones of non-contracting parties. As the French delegate explained: 'Accession implied certain obligations, as well as certain measures of protection, and it seemed illogical that a state which did not recognise the obligations of the convention in regard to its own ships should, nevertheless, enjoy all the measures of protection.' On the next day, the French proposal was appended to a British proposal for a future conference to consider reducing the size of zones for non-contracting parties.

But the attempt to enshrine a principle of reciprocity in the amended convention was thwarted by the strong opposition of delegates at the Legal Committee. The American delegate said that they preferred to act by way of encouragement rather than by way of threats. The Soviet delegate also pointed out that the principle of reciprocity would be contrary to the humanitarian, legal and political merits of the pollution convention. When the French rider to the British proposal was defeated by a large majority at the legal committee, the French delegate, Jean Roullier (who later served as IMCO Secretary-General), noted the potential consequences of their decision, thus:

> At the Conference on the Safety of Life at Sea it had been decided that countries which did not ratify the convention would not be treated like the others, and therefore their ships were subject to a supervision which was different from that imposed on the ships of the contracting states . . .
>
> The rejection of the French proposal meant that the situation would be reversed: not only would the ships of Contracting States [to the oil pollution convention] not be treated better than the others; they might even be less well treated. If a ship flying the flag of a Contracting State was suspected of having infringed the regulations, a report would be made to the government whose flag the ship was flying, but in the case of a ship of a non-Contracting State, no measure of that kind would be taken.

> . . . it therefore seemed that it would be in the interests of States not to ratify the Convention, as in that way they could enjoy its protection without submitting to its regulations.

The zones protecting eastern Canada, Iceland, Kuwait, Norway and countries bordering the Mediterranean and Adriatic were extended to 100 miles. The Australian zone was extended to 150 miles from the nearest land, except in the north and west coasts of the Australian mainland where a wider zone applied. The Atlantic zones off the eastern part of the U.S., France, Ireland and Britain were also extended, the latter now obtaining a limit of 40° west longitude for greater protection. The Baltic and North Seas were closed entirely.

For countries which had not ratified the 1954 convention, the following extended zones and closed areas were granted after their acceptance of the amended treaty: 100 miles for Portugal, Spain, Saudi Arabia, Malagasy and countries bordering the Red Sea; 100 miles for countries bordering the Black Sea and Sea of Azov, and total prohibition after the Soviet Union and Rumania ratified the treaty; a special zone for India in the Bay of Bengal, Arabian Sea and the Indian Ocean.

Closure of the North Sea

Discussions surrounding total closure of the North Sea to oil pollution were particularly reflective of the yet sensitive regard of some countries to zones where they had extensive shipping interests, as well as the significant advances made by the 1962 conference. The 'hole' left in the North Sea by the 1954 convention was identified as the source of serious pollution in Britain, West Germany, Denmark and Sweden, except Norway where prevailing winds countered the drift of oily spills.

Britain led the others in asking for total closure of the North Sea, but at the same time tabled an exemption for coasters under 500 tons which operated in the area, since it was held that these coasters would not have been able to meet the prohibition requirements when port facilities were not considered adequate for their needs. On the other hand, Norway alone resisted total closure of the North Sea and inclined towards a 100-mile zone from all surrounding countries. Captain Neuberth Wie (Norway) explained that 'drastic steps were not wanted, and it was preferable to extend the coastal zone to 100 miles and retain the gap.' Moreover, he pointed out that reports of oil pollution originating from the North Sea might be due to ships of non-convention countries. But everyone in the zones committee, including its chairman (Captain C. Moolenburgh of the Netherlands) argued for total closure of the gap. Captain Wie insisted that Norway precisely needed the hole in the North Sea, not only for their own ships, but also for ships of other countries, to wash out their tanks. The negotiations turned a corner when Lord Hurcomb of the British delegation made a personal plea for the protection of wildlife:

> Birds know no international frontier but only find to their cost that, if they came down where the sea was polluted by oil, they were cruelly destroyed. The bird population, particularly the most rare, beautiful and harmless species, had been decimated in past years.

The Norwegian delegate asked to confer briefly with other members of his delegation, and, as Lord Hurcomb later described, 'quite fairly threw' the matter back to the British delegation.[234] Captain Wie replied:

> Norway had a considerable merchant fleet and a large part of it was tankers. The Norwegian delegation had been rather concerned that shore facilities might still not be sufficient in the North Sea, but to extend the 50-mile zone to 100 miles along the whole of the Norwegian coast [and that of the other countries]. However, in spite of the inconvenience, the delegation had intended to go as far as to accept closing the gap on the condition that it should not come into force until a certain number of States, for instance at least thirty countries, including all countries with not less than one million gross tons of shipping, had accepted the Convention. However, after having heard the state-

> ment of the distinguished United Kingdom delegate who spoke so stirringly and seriously about sea-birds, the Norwegian delegation had decided not only to agree to the closing of the gap, but to go even further, and prohibit also the small ships, which would not be covered by the amended Convention, from pumping out oil-contaminated water and tank-washings into the North Sea, as it was believed that oil-contaminated water would be just as dangerous for the birds whether it was pumped from smaller or larger ships.

This exchange caused some consternation when the other British delegates, at this stage anyway, refused to concede. Mr. Haselgrove (UK) said that:

> a large number of very small ships operated in that area, and until adequate facilities provided all round the North Sea, it might not be practical to apply the Convention in all its rigour. His delegation had favoured the retention of that exemption because some countries might find it difficult to accept the Convention without it when they came to examine its implications in practice.

Mr. Gillender (UK) added that without the exemption a large number of North Sea coasters would have to detour until adequate port facilities for them could be provided. Captain Wie of Norway replied that 'if Lord Hurcomb was satisfied with the United Kingdom position as expressed by Mr. Gillender, he too was ready to accept it.'

The American delegate did not find the exemption for small coasters convincing. In this event, the British delegation found it quite embarrassing to keep to their previous position, and as a result of backstage consultations, they withdrew their proposal to exempt coasters on the next day. The North Sea was declared a special area, to be totally closed to oil dumping by ships. This legal status prevailed until the 1973 Marine Pollution Convention was drafted, when the North Sea became the only special area to lose such a privileged protection from oil pollution.[235]

'Total Prohibition' Achieved

The advances in pollution control gained at the 1962

conference pale in significance to what many delegates considered as the greatest achievement at the time – total prohibition of oily discharges from all new ships of 20,000 gross tons or more. The British delegation had resurrected another version of their previous proposal given at the 1954 conference, except that this time they wanted to require new tankers and non-tankers of 20,000 gross tons or more to retain their oily ballast and cleaning water onboard, subject only to exception when the ship in question sailed between destinations without port facilities, in which case the ship could wash out only outside the prohibited zones.

The 'no discharge' criterion for new large ships was also intended to apply eventually to other ships and tankers already in service, after a grace period. As the British delegate pointed out, their proposal was 'the most far-reaching of all the proposals before the committee, as it aimed at progressive improvement of the position.' In technical and economic implications, it also proved to be the most controversial motion before the 1962 conference, even though it was by no means a new idea.

A special working group was assigned to study the matter at the conference over the opposition by the U.S. delegate who wanted the proposal considered by IMCO at a later time. The group was informed of successful sea voyages with total tanker retention of oily wastes by a British company.[236] Several formidable obstacles to the procedure were analysed. Tankers could continue to discharge their oily ballast and cleaning water at sea, provided they retained and separated oily slops onboard in a special slop tank. Short voyages, inclement weather, and human error could prevent the adequate separation of the oil from the ship's ballast or cleaning water. Secondly, the slops of the old cargo had to be compatible with a new cargo of oil or otherwise acceptable to a refinery or terminal Thirdly, title to the oily slops would be questionable if the tanker belonged to an independent charterer or competitor. Fourthly, the oily slops would displace some amount of prime oil for which the operator (if the slops were discharged at a terminal) might have to face penalty in loss of deadweight tonnage and carriage of dead freight. Fifthly, the operator might have to pay customs dues on the oil slops, increased harbour fees for cleaning, and tonnage calculation dues. And finally, Suez and Panama Canal authorities required tankers to enter their canals free of oily ballast or slops ('gas free') for safety reasons, or otherwise charged them fees equivalent to a fully-laden tanker.

The American delegation led others in counselling against the wisdom of implementing total tanker retention of oily residues onboard, even as they indicated their desire to attain such perfection at some future date. In their view, total prohibition for new large ships at this stage seemed impractical due to the lack of general experience and acceptance for such procedures by the oil and maritime industry. They circulated their own memoranda to apprise the delegates of the precipitousness and technical anomalies in the British proposal. The technical and administrative problems connected with the prohibition could best be dealt with after more more careful study by IMCO, in the view of the opponents to immediate requirements. As James Moss (US) noted, the 'substitution of panic action for studied action' could hardly do justice to the interests involved. Secondly, the American delegates questioned the capability of tankers to retain slops and discharge only such oily mixtures that would be within the permissible limits of the amended convention. In their view, the procedure 'cannot be accomplished under the convention definition of oil pollution without violation,' and the memorandum of the British tanker company conceded this. Indeed, the relevant passage in the British tanker company memorandum stated that it was not possible to measure the content of oil in the oily mixtures onboard. The colour of the discharge, under their suggested procedure, would yield an effluent which changed rapidly from nearly colourless liquid to 'brown' when the oil contamination in the discharge became appreciable.

On this technical anomaly in the British proposal, the U.S. delegation admonished that it would cause the conference 'to temporise with the very foundation piece of the convention – namely the definition of oil pollution itself.' As a helpful gesture, the U.S. delegation offered that they should consider changing the definition of what constituted oil pollution from '100 parts per million of oil' to 'a brownish liquid which will finally disappear', and this would accurately reflect what was being proposed. Alternatively, they suggested placing 20,000-ton tankers or larger tankers in a category not covered by the oil pollution limit, a 'situation [which] would still be discriminatory but it would at least be factual.' If the conference accepted the new requirement on tanker discharge prohibition, then the U.S. delegation warned that they could injure the credibility of the amended convention and be regarded 'by tanker men the world over as rather irresponsible' in their work.

However, the British delegation fielded their best men and their own technical experts to tip the advantage in their favour at the working group stage. Moreover, they modified their original proposal to accommodate the objections raised by the other delegations.

The working group reported out a modified British proposal, and the committee on ships proceeded to discuss the motion. At this stage, the U.S. delegation could have formally introduced their recommendation to change the definition of oil and to place new large tankers in a special category altogether. Their failure to do so was later explained by one former delegate as futile – 'there was little interest shown by others in these alternatives.'[237]

The committee on ships accepted a revised British draft which attracted more comments by other delegations, and further concessions were given to accommodate these. The U.S. delegation repeated their objections and warned that they might lose the acceptance of oil and maritime interests if the proposal was accepted, thus implying that the U.S. would not ratify the amended convention. The Japanese delegation repeated their wish to shunt the 'no discharge' requirement on new large tankers into a recommendation rather than a binding obligation on contracting parties, which corresponded with the positions of the Netherlands and Norway. But undoubtedly due to the insistence and influence of the British delegation and their technical advisers, as well as a promise from Suez Canal authorities to alter their policy on oily ballast dues (a concession achieved through the personal intervention of Sir Gilmour Jenkins as conference President), the majority favoured approval of the British proposal as a binding obligation in the amended treaty. A few more changes were made, and the British proposal for total prohibition of oily discharges from new large tankers and non-tankers of 20,000 tons or more was passed by the conference.

At the final plenary session, Sir Gilmour summed up the achievements of the 1962 conference thus:

> First. We have extended the Convention to cover more classes of ships than before. All ships down to small ones of 150 tons gross tonnage have been brought completely within the Convention.
>
> Second. Governments have also agreed to apply the provisions of the Convention as far as is reasonable and practical to all their ships of whatever size and also to their naval vessels.
>
> Third. The principle that ships shall in no circumstances discharge persistent oil into the sea has been accepted. We have gone no further at this stage than to apply the principle to all new vessels of 20,000 tons gross tonnage or more. Even here we have had to provide for exemption when reception facilities ashore are not available at either end of the voyage. Full details of such discharges have to be reported to IMCO. This provision is a sign-post to the future and our Resolution No. 4 urges that all tankers even those of less than 20,000 tons gross tonnage should follow suit where this is reasonably practicable. We have been told that some tanker companies are already operating their tankers in this way. I hope that more will follow their example.
>
> The fourth achievement of the Conference has been to extend the requirement [sic] that reception facilities should be made available for waste oil from dry cargo ships to cover facilities at ship repair ports and for tankers at oil loading terminals as well. It has also been amended to make it easier for certain countries to accept it . . .
>
> The fifth main achievement of the Conference – and a more important one – is to have made very great increases in the extent of the zones in which oil must not be discharged into the sea . . .

But the 'radical cure' for the oil pollution problem yet eluded the 1962 conference, as Sir Gilmour admitted:

> ships should never, in any circumstances, [intentionally]discharge any oily waste into the sea. This means that there must be facilities ashore into which ships can discharge them.
>
> We come back then to one measure, well within the capacity of all countries of the world and of the . . . oil companies . . .

The 1962 Amended Convention

The International Convention for the Prevention of Pollution of the Sea by Oil, 1954, as amended in 1962, came into force on 18 May 1967, only two months after the *Torrey Canyon* disaster off Cornwall, England. As such, the 1962 amended convention could have served as the lynchpin of the search for improved oil pollution control measures. However, due to the introduction of the Load on Top system by the major oil companies, and subsequent legitimisation of this system by the 1969 amendments, the 1962 amended Convention became an anachronism before it had even been given a chance to work. Our analysis of the amended treaty will thus include only its more notable features which influenced later events.

Although the delegates to the 1962 conference understandly felt that they had achieved much, their advances seemed credible only as good recommendations for governments and the oil and maritime industry, but hardly effective without the subsequent support of the latter. Where the 1962 amended treaty represented a potentially strong deterrent against shipping pollution, its successful application depended on a confluence of factors beyond the scope of agreement.

IMCO and the 1962 Amended Convention

The 1962 amended convention enlarged the scope of IMCO's role in oil pollution control and maritime affairs in general. Since then, this extension of IMCO's functions has made the organization (now called IMO or the International

Maritime Organization) the pre-eminent international agency for the discussion and co-ordination of global maritime policy and practice, especially for safety in navigation and control of shipping pollution. Indeed, after the spectacular pollution disasters of the 1960s and 1970s, governments generally have chosen to negotiate related problems of controlling pollution either within the IMCO structure or through IMCO-sponsored international conferences.

IMCO as a central agency for pollution control issues has its potential uses, especially for the maritime states which (then) mostly controlled the agency's policy-making committees. As the central agency envisaged since the 1920s, IMCO took over these functions from the British Government on 15 June 1959. The 1962 amended convention and other resolutions of the 1962 conference expanded IMCO's role, despite the troubled early history of the organization. As the depository of oil pollution conventions, IMCO accepts and communicates ratifications and accessions, denunciations, extensions of the rules to colonies, and suspensions of the treaties by state-parties. IMCO prepares and updates the charts of prohibited zones.[238] IMCO can also receive reports from state-parties on discharges within prohibited zones and details of penalties imposed for convention ships.

Apart from receiving and distributing information on oil pollution topics, IMCO has two more important roles to play. Firstly, IMCO can be used an agitation forum to discuss and to spur government action on vessel-source pollution control, even though in later years many United Nations agencies and international organizations have become interested in environmental problems. Related to this function is IMCO's pliability as a means of encouraging accessions and as a source of information and technical assistance.

Secondly, when a diplomatic conference and unanimous agreement on changes to the treaty were not possible, IMCO has been used to put forward amendments to the oil pollution treaty, for example in 1969 and 1971, through the adoption of these amendments by the IMCO assembly which meets every other year.

Balance of Obligations

But enforcement and implementation of international treaties remain with state-parties. In many ways, the oil and shipping industry carried a larger burden of responsibility than coastal authorities. State-parties did not have to provide port reception facilities, nor did states in general have to accept the treaty to enjoy a zone of protection off their coasts.

In order to make the amended treaty more acceptable, state-parties were relinquished of their previous obligation to 'ensure' the provision of reception facilities for non-tanker oily wastes. The new Article VIII was worded in such a way as to invite state-parties to 'promote and encourage' port authorities, which may be independent bodies, or oil company terminals, or shipyard repair ports, to provide the necessary receptacles. If these shore facilities were found to be inadequate or non-existent, it was entirely possible for state-parties to plead that these facilities were beyond their jurisdiction or resources, and that in its formal sense, Article VIII only asked them to 'take all appropriate steps' to promote their provision.

Furthermore, except for the resolutions calling on governments and canal authorities to remove various requirements which disabled the collection of spent lubricating oil and the passage of tankers in oily ballast through international waterways, the 1962 amended convention did not explicitly oblige state-parties to refrain from the collection of taxes on oily slops or to charge for the use of reception facilities. Thus, shipowners or tanker operators who desired to retain the oily wastes onboard and bring these to port, in many cases, would have suffered financially because of the policy of coastal authorities whose coasts they were presumably trying to protect in the first place.

But convention ships had to reconsider their use of the seas as a vast sink for oily wastes. Oil-fueled ships (non-tankers) yet using their fuel tanks for ballast, the main source of pollution in non-tankers, were required to refrain from washing out at sea if they proceeded to a port with reception facilities. Both tankers and non-tankers were required to observe greatly expanded prohibited zones, especially where such ships congregated.

This surfeit of obligations upon shipping, however, appeared greatly reduced by various qualifications in the provisions of the convention and also by the lack of an effective machinery for enforcement. Coastal state parties had no powers of arrest or detention over a suspected vessel for offences in prohibited zones, and final authority continued to rest with the flag state. Flag state jurisdiction, as we have seen, opened the door to a number of abuses and anomalies even if the ship belonged to a state-party. The most that a coastal state-party could require of convention ships was to examine their oil record book, and report any violation to the flag authority. Even certain detection by means of sightings at sea of illegal discharges within zones or by classes of ships prohibited from such discharges, could only be reported back to the flag state party.

Conditional Ban on Pollution

Overall, the 1962 amended convention brought the idyllic aim of total prohibition closer to realisation by means of the conditional ban on mid-oceanic discharges by new large tankers and ships. Thus, Article III (c) of the treaty stated:

> . . . the discharge from a ship of 20,000 tons gross tonnage or more, to which the present Convention applies and for which the building contract is placed on or after the date on which this provision comes into force [18 May 1967], of oil or oily mixtures shall be prohibited. However, if in the opinion of the master, special circumstances make it neither reasonable not practicable to retain the oil or oily mixture on board, it may be discharged outside the prohibited zones . . . The reasons for such discharge shall be reported to the [flag state]. Full details of such discharges shall be reported to [IMCO] at least every twelve months by Contracting Governments.

Departures from the above norm appeared to have allowed only exemptions due to emergency and safety, allowances for ships travelling between ports without suitable reception facilities, independent operators whose oily wastes were not acceptable to major oil companies, and ships liable for other disabilities due to canal regulations, port tolls and customs dues on oily ballast. The procedure for departing from total prohibition in new large ships represented, at least to some advocates, a considerable hedge against abuse. Each departure had to be reported and justified to the flag authority. Upon certifying the exemption, the state-party had to render an annual report to IMCO of these cases, rendering them liable to public censure for the actions of their vessels.

Shipmasters would have assumed the burden of proof for departing from this conditional ban on new large ships. All that a loading authority had to do was to ascertain that the ship contained the apporpriate rate of oil remaining in tanks. Completely clean tanks would have led to the presumptive conclusion that the ship in question had cleaned somewhere at sea. Also, the loading sheet of oil company inspectors who monitor the amount of oil loaded into tankers would have constituted prima facie evidence against a tanker with clean tanks, which would have taken in more oil than a tanker with some retained slops onboard.

Different Paths to Attain the Aim

As the 1962 amended convention attracted ratifications, research escalated on ways by which new large ships could meet the conditional ban requirements. As the American delegate told the 1962 conference, they were willing to abide by total prohibition 'possibly by a rather different road.' In effect, American interests intensified the development of segregated ballast requirements for new tankers.[239] Merchant ships operated by the Soviet Union opted for another antipollution practice, using chemical solvents in a closed system of tank cleaning and recovery of oil residues.[240] Meanwhile, British interests led another promising area of research into what later became the wider practice of the Load on Top system of tankers retaining oily slops onboard and mixing these with a new cargo of oil.

These inquiries came at a time when significant developments in the oil and shipping industry made it necessary to adopt suitable practices in keeping with the affirmed objective of protecting the marine environment from pollution by ships. By the mid-1960s a tanker of 20,000 gross tons had become a midget as compared to the size of new or converted Very Large Crude Carrier (VLCC) tankers of 200,000 tons or more carrying capacity, which became the popular means of bulk transport. Actual oil demand outstripped forecasts of five per cent actual growth, and world shipping fleets expanded in line with booming world trade and seaborne movement of oil and other cargo.

Total closure of the Suez Canal after the June 1967 war vindicated a major transition from medium-size carriers to the VLCCs, thus effectively doubling the size of tankers over a period of two decades. Moreover, the economies of giant tankers were said to be 'irresistible' when it cost half again as much to use a 200,000-ton tanker as a 50,000-ton tanker to transport Middle East crude to Europe through the extended journey round the Cape of Good Hope.

As tanker sizes and fleets expanded, it became increasingly difficult to accommodate passage of these superships through existing routes and their entry into ports. Very few terminals could handle the VLCCs, and new ports could not be built overnight. Noval requirements of land and water had to be met since giant tankers need sufficient depth of water to enable them to bring their full load to port and an area of about 500 acres to handle storage tanks or new refineries. Offshore moorings had to be re-designed to handle the VLCCs, and plans began for the construction of new oil terminals in Western Europe.

There seemed every reason to suppose that progress in meeting total prohibition for new large ships would have exerted considerable pressure on existing ships to follow suit. Indeed, it was generally assumed that the industry could meet the requirements for the conditional ban against oil pollution, and that another international conference would be called in a few years time to revise the 1962 convention accordingly.[241]

The oil and maritime industry acknowledged that continued or worse pollution could easily have stirred public opprobium and movements for unilateral measures by aggrieved coastal states. As time has shown, massive oil pollution disasters caused by shipping casualties, on top of recurrent intentional pollution by ships, fueled the very anxieties which the industry had wished to avoid.

By the mid-1960s, the major oil companies co-ordinated their tactics and announced that they had 'discovered' a system which was the 'cure of operational pollution.' After the first public announcement of their new strategy, a massive publicity campaign was launched, which for its effect must be one of the most successful propaganda stories of all time.

Load on Top and the 1969 IMCO Amendments

On 17 June 1964, three of the world's largest oil companies, Shell, British Petroleum and Esso, which between them own or charter sixty per cent of the world tanker fleet, announced that after three years of trial and research they had come to the conclusion that the practice mainly responsible for causing oil pollution by ships – the cleaning out of all empty cargo tanks – was unnecessary for most purposes.

The origins of Load on Top (LOT) were described more fully by John H. Kirby, Director of Shell Marine International, to the 1968 Rome conference of interested parties:[242]

> The outcome of the 1962 Conference was a source of great disappointment and grave concern to a great many interested people and notably to one of the world's leading ornithologists . . . Lord Hurcomb who expressed his feelings to some of his colleagues in the House of Lords, including the then Chairman of my parent company. As a result, I was asked. whether, notwithstanding all the advice that Shell and our major oil company contemporaries had contributed to the 1962 meetings, we could

> not take an entirely new look at the problem and develop and evolve a real solution . . .
>
> The tanker industry people involved in providing advice for the 1954 and 1962 Conferences (including representatives from Shell!) still held the view that it was necessary to have a clean container (. . . tanker) and as free of sea water as possible for the next cargo, because that had always been expected of tanker owners by their customers, i.e. the Oil Industry. But sparked off by Lord Hurcomb's question, we in Shell put on our oil industry hat rather than our tanker-owning hat and asked ourselves the question: Do we really need a clean and dry tanker for every cargo in the crude trade?
>
> Although the instinctive first answer based on the practice of the trade for years was still affirmative, we arranged for laboratory experiments. The results of this were inconclusive but tended to reaffirm the status quo. We refused to accept this and insistence on a full-scale experimentation produced the long sought answer of *no* – we do not need a clean and dry tanker for every cargo in the crude trade!
>
> This was the breakthrough. Most crude oil cargoes are compatible with one another within the limits required for Load-on-Top operations and the little extra salt can be accepted without major difficulty or cost in the refining process.

In future, these major oil companies agreed to pump out only that part of oily ballast from which most oil residues had been separated and retained onboard as dregs of the previous cargo, mixing these with the next cargo of crude. A few cargo tanks would still be washed for the operational efficiency of the tanker. See Fig. 3 for a pictorial diagram of the LOT method.

It was said that generally the traces of oil in LOT discharges would be well below 100 ppm, but as the oil layer was reached, the discharges gradually turn darker and may exceed 10,000 ppm. When the shipmaster notices the darkening discharges, he was instructed to cut off pumping and to transfer the remaining slops to a tank designed to receive the contaminated mixtures. As much oil and sea water emulsions are retained onboard the tanker, and the next load of crude is simply loaded on top of the previous slops.

Figure 3. The Load on Top System

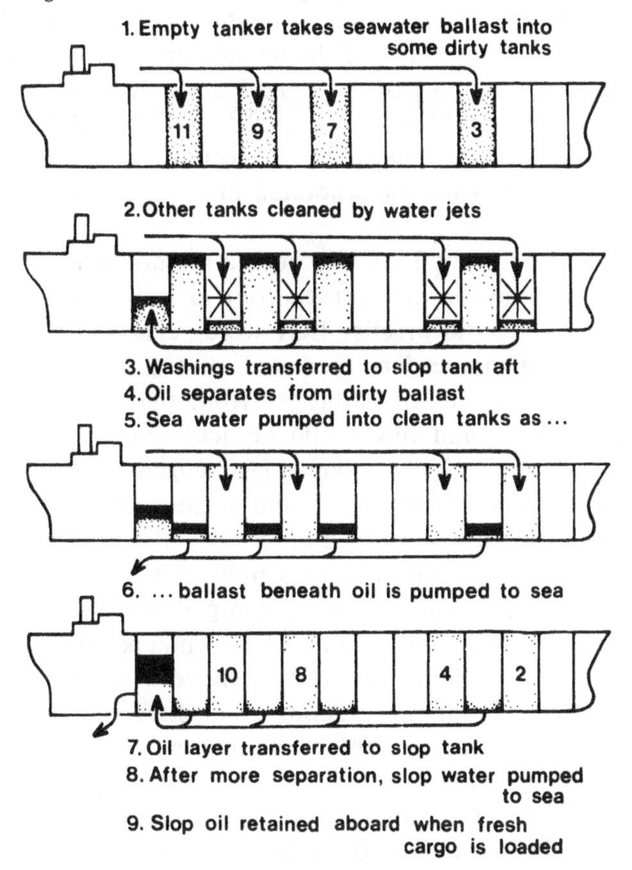

Source: A Nelson-Smith, Oil Pollution and Marine Ecology (London: Paul Elek Ltd., 1972).

Secondly, the ship's discharge would be pumped out to sea more slowly than the traditional method of massive pumping during short periods. This would allow the oily effluent to pass through the propeller stream and be dispersed in the wake of the ship.

And, finally, when added to the next crude load and brought to a refinery, the LOT mixtures constitute a 'novel' type of crude in two ways: they have an enhanced wax content, and they contain some sea water. A tanker bringing in a cargo of LOT dregs in the slop tank would normally have been turned back by refineries because of the novel mixtures which it brought. Thus, the ultimate acceptance of LOT hinged on the refining arm of the oil industry which eventually came to the support of the tanker transport departments. Major teething problems were solved, and the LOT tanker practice, combined with refinery acceptance of LOT slop tank mixtures, became incorporated in an oil industry 'Clean Seas Code.'

It appeared that LOT was both a method leading to reduction of pollution and conservation of oily wastes. If all tankers which could practise LOT were persuaded to do so, then oil company leaders predicted that LOT would cure 'ninety-nine per cent of pollution' from the normal operations of tankers, and this 'could be achieved in a relatively short time.' The major oil companies pledged to achieve universal acceptance of LOT by all the means at their disposal. Based on 1967 statistics of oil movement at sea, a 'pollution profit and loss account' was presented to show how much LOT had prevented oil from being dumped at sea in that year alone: of the total crude shipped in tankers (700 million tons), 2.1 million would have been dumped without the LOT practice, but the voluntary adoption of LOT by eighty per cent of oil companies at 99% efficiency was claimed to have saved 1,663,200 tons of oil from being discharged at sea, leaving only the LOT discharges 'in-polluting form' (16,800 tons) and the discharges from tankers not practising LOT (420,000 tons).[243]

One of the most significant drawbacks to the implementation of LOT was said to be the 1962 amended Convention itself. Under the definition of pollution (100 ppm) in the treaty, LOT was supposed to be an illegal procedure because it could lead, especially during the last stages of pumping, to discharges exceeding the permissible limit.

Therefore, the major oil companies successfully advised the revision of the 1962 amended Convention in order to legalise LOT. Officials and environmentalists became overwhelmingly convinced of the merits in the new practice, and they willingly changed the 1962 regulations. On 21 October 1969, the 6th IMCO Assembly passed amendments to the

1962 treaty in order to legalise the practice of LOT.[244] Among the 1969 amendments were the following provisions: (1) the new meaning assigned to 'oily mixture' would be a 'mixture with any oil content' rather than the 1954/62 limit of 100 ppm; (2) existing requirements for tanker observance of prohibited zones were deleted, and the new procedure for tanker washings permitted a tanker more than 50 miles from land to discharge oily wastes, provided that the rate of discharge from the tanker does not exceed 60 litres per mile of travel; (3) to guard against unnecessarily prolonged discharge of oily wastes, the total amount of oil discharged was limited to not more than 1/15,000 of the total quantity of the original cargo; and (4) a new form of oil record book was appended to the 1969 amendments to reflect the approved practice of LOT.

A Step Backward

However, the 1969 amendments also applied LOT to new large tankers, thus deleting the 1962 requirements for the observance of total prohibition against pollution by new large ships built after 1967. When the 1969 IMCO amendments were first made public, the omission seemed less important than the desire of the oil majors to convince the public of the advantages of LOT. After strong criticism, however, a lively debate in specialised circles ensued between some environmentalists, led by Counsellor Boos, and the LOT advocates, led by Shell Oil company spokesman.[245]

The controversy involved both the meaning and practical effect of the deleted clause, Article III (c) of the 1962 amended Convention. LOT advocates explained that the 1962 clause did not really require absolute prohibition against oil pollution by new large ships, because discharges of less than 100 ppm would still have been allowed. Against this argument, Counsellor Boos held that the 1962 provision would have effectively banned any oil discharge, since tankers would have had to retain oily mixtures onboard in order to stay within the convention. As the records of the 1962 conference showed, the American delegation did warn that the intended effect of the new clause was to prevent tankers from discharging oily wastes at sea.[246]

Secondly, the defenders of the deletion noted that the 1962 clause would not really have imposed total prohibition on routine pollution when the shipmasters could decide that 'special circumstances' made it neither reasonable nor practicable to retain oil or oily mixtures onboard. However, the discretion to masters was strictly limited, as Counsellor Boos explained:

> It is granted as a special exemption right to the master, who further is obliged to report to his government on each occasion when he takes advantage of the concession, giving the reasons for his action. Thus, the concession cannot reasonably be represented as depriving the clause III (c) of its main import.

Finally, it was also alleged by the oil companies that new large ships would not have been permitted by technology to meet the total requirements of the 1962 amended Convention. On the contrary, Counsellor Boos replied that total avoidance of oily discharge from ships was feasible even then.

The formulators of the 1969 amendments maintained that they did not view the deletion of the total ban on pollution by new large ships as a retrogression in practical terms. However, to Counsellor Boos and others, the decision was not only 'a deplorable step backward,' it was also 'a grave mistake' to remove the 'signpost of the future' from the oil pollution convention.

6

AFTERMATH OF THE *TORREY CANYON* DISASTER

Due to the proliferation of tanker disasters resulting in massive coastal pollution in the 1960s and 1970s, governments and private interests became more inclined to face problems which had been ignored before – the problems of preventing accidental pollution by ships and of compensation and remedial measures for oil pollution damage. The *Torrey Canyon* disaster made pollution a highly topical problem, with subsequent tanker accidents sustaining public interest in control measures, that the history of pollution control turned into a series of attempts to deal with 'pollution disasters'.[247]

Although the amount of leakages of oil from accidents pale in comparison to the annual levels of deliberate pollution by ships, nonetheless, accidental spills occur visibly closer to economically productive coastal regions and articulate communities, and thus appear more spectacular. Such environmental consciousness since the 1960s contrasts sharply with the relative lethargy of earlier decades. Whilst before, the problem of oil pollution was the specialised concern of a few leading countries, a few major oil companies, and conservationists, the tanker casualties created a democratisation of policy-making, enabling more countries and a wider public to contribute to the formulation of policy.

Secondly, the popularisation of oil pollution control problems has resulted in wider publicity and dissemination of information on the subject. Today, few environmental problems have equalled the extent to which oil pollution has been popularly portrayed, seriously studied, and officially dealt with.

Greater public consciousness about environmental problems involving air, land and water quality stirred up nothing short of an 'environmental revelation' which has led to the creation of new international organizations (e.g. the United Nations Environment Program) and new international agreements and regional arrangements, inducing national governments to include pollution prevention policies as a matter for national interest.[248] For purposes of the present study, we shall only briefly recount the *Torrey Canyon* incident, the most influential maritime casualty in recent history, and survey some of the advances in oil pollution control measures taken since then.

The *Torrey Canyon* Disaster

In the early morning of Saturday, 18 March 1967, *Torrey Canyon,* then the thirteenth largest merchant vessel, was steering a dangerous course on automatic pilot off the southern coast of Britain, between the Seven Stones Reef and the Isles of Scilly. In the oil transport trade, *Torrey Canyon* was considered a prime asset, but in years to come such types of tankers would appear as 'modern pirates' to coastal residents or visitors who have shared traumatic memories in the wake of tanker pollution.

The *Torrey Canyon* incident would involve many states. While originally built in the United States in 1959, the tanker was 'jumboised' in Japan in 1964. It was registered in Monrovia and flew the Liberian flag, although it had never been to Liberia. It was owned by the Barracuda Tanker Company which maintained 'filing cabinet' offices in Hamilton, Bermuda, and in Monrovia, Liberia, but was essentially a transport arm of the Union Oil Company of California. The officers and crew were Italian. On this fateful voyage, the tanker had been chartered from Union Oil by British Petroleum, laden with over 100,000 tons of Kuwait crude oil and bound for their refinery at Milford Haven, Wales. *Torrey Canyon* was considered to be excellently equipped and manned – it carried the highest seaworthiness rating of Lloyd's register of shipping.

It was only in the ensuing disaster that a more searching scrutiny would show that, despite the best crew and the best-designed and equipped ship, accidents do happen. Captain Pastrengo Rugiati had received instructions from the demi-charterers to catch the high tide at Milford Haven to allow his over-50-feet draught tanker to clear the harbour. During the last critical moments, Captain Rugiati desperately tried to control the steering but he could not turn the tanker easily, and *Torrey Canyon,* at full speed and in broad daylight, rammed Pollard Rock, part of the Seven Stones. The grounding took place approximately eleven miles off Britain's toe, which was then outside British territorial waters. Immediately, a large breach in the hull of the tanker resulted in massive leakage of oil to the sea. The action of wind, tide and currents eventually brought the oil, and its ugly consequences, to Britain and, nearly a hundred miles away, to France.

The British and French governments faced their first major oil pollution crisis with little or no preparation at all. In this instance, remedial action was determined largely by events rather than by policy. Emergency measures were put in hand to cope with the oil and the wounded tanker. Royal Navy tugs began to spray detergent in an effort to disperse the oil at sea. A private salvage tug of the Dutch Bureau Wijsmuller garnered the contract of trying to salvage the tanker on the basis of Lloyd's 'no cure no pay' salvage contract. Although immediately at risk, Britain had no powers to deal with the ship until or unless the owners and their appointed salvors had decided to abandon the vessel.

Meanwhile, in Whitehall, emergency cabinet meetings discussed the requirements of the situation as they arose. Various ministries were assigned emergency duties, and local authorities in the endangered areas were assured of government assistance and partial subsidy in their efforts to prevent and to mitigate the pollution. The British government had three options – to hope for the salvage of the tanker; to save the cargo; or to burn the oil in the tanker where it lay. It has since been recognised that the best course at once might have been the second alternative – to recover the oil before it caused more damage – since the first course, though grounded in tradition and law of the sea, was not necessarily the best means of reducing pollution from a large tanker casualty, and the third (bombing a tanker) was regarded as 'the very last method of despair.'[249] In view of the novelty and difficulties of the situation, however, the British government was forced to take calculated risks. Interestingly enough, they had refused an earlier offer from the salvage company to enter as a sub-contractor in the salvage effort.[250]

The difficulties of salvaging the tanker were compounded by the death of the salvor leader on 21 March, the breaking apart of the tanker on 27 March, and the separate decisions of the British and Dutch governments to deny the salvage company access to one of their ports in case salvage had been successful. On 25 March, the first slicks of oil arrived at Cornish beaches, and the Labour government of Prime

Figure 4. The Torrey Canyon Disaster

Source: IMO News No. 1:1984

Minister Harold Wilson met the biggest home-front emergency since it took office.

When salvage efforts were finally abandoned, the British government decided that the time had come to burn the remaining oil in the tanker in order to control pollution. On 28 to 30 March, Royal Navy planes repeatedly bombed the tanker with incendiaries and explosives, and the *Torrey Canyon* became a flaming mass which sunk into the sea – at the time the most expensive maritime casualty (see Fig. 4). The loss of the ship itself cost insurers $16.5 million and nearly another $1 million for the loss of the oil cargo. British and French clean-up costs amounted to between $14-16 million total.[251]

Rarely has a maritime incident had such widespread and immediate effects. There followed a great deal of speculation, inquiries, court action and litigation to determine the cause of the tragedy and to apportion the blame. The British government took their fair share of criticism for the hesitancy of the action which was finally taken to set fire to the oil remaining in the tanker.[252] The Liberian government, smarting from innuendoes that *Torrey Canyon* was of inferior quality, being a flag-of-convenience ship, conducted their own inquiry at Genoa, Italy, where the crew had been hired, and a summary of the Liberian investigation in May 1967 placed the entire blame for the disaster on the shipmaster rather than to deficiencies of the vessel or its owners.[253]

In the wake of the *Torrey Canyon* and dozens of other coastal pollution incidents involving large tankers throughout the world in the 1960's and 1970's notablv that of the *Amoco Cadiz,* which grounded off the coast of Brittany, France, on 17 March 1978 resulting in a more extensive pollution of the French coast and more expensive clean-up costs and damage claims amounting to well over $100 million, a momentum was sustained for national and international action on accidental pollution by ships. The seriousness of tanker casualties in the 1960s-1970s involving tankers of over 500 deadweight tons capacity had more than doubled since than 1950s. A U.S. Coast Guard survey in 1971 showed that in one sample year alone there was at least one pollution incident by tanker accident in every five days, and that most of the accidents occurred either near to shore or at port.[254] As the British parliamentary report on coastal pollution concluded, 'The real lesson of the *Torrey Canyon* is that in an operation of this kind involving this kind of quantity of oil the risk and subsequent expense arising from oil pollution should be given a greater weight earlier in the proceedings than perhaps it was.'[255]

At present there are some thirty conventions and other types of international agreements relating to maritime safety and the prevention of pollution by ships. We shall briefly survey these measures against accidental pollution by ships under the following topics: (1) prevention of accidental pollution; (2) coastal state intervention; (3) liability and compensation for pollution damage by ships; and (4) mitigation and clean-up measures.

Prevention of Accidental Pollution by Ships

Prevention being more desirable than post facto measures, the following improvements have since been made to international regulations on maritime safety and pollûtion control – better design and construction of tankers; amendments to the Safety of Life at Sea convention; the entry into force of the international loadlines regulations; traffic separation and routeing schemes throughout the world; and better training and certification of mariners.

Changes in Tanker Design and Construction

The *Torrey Canyon* disaster came at an historic turning point in the life of oil tankers when the average size of tankers in use was about to double. As we have seen, the economies of transporting oil by means of VLCCs were so irresistible that plans had been made for the construction of and use of even larger carriers in the 500,000 deadweight-ton capacity or more (Ultra Large Crude Carriers or ULCCs). However, the 1971 IMCO Assembly agreed to regulate the future design and construction of tank sizes in tankers in order to prevent massive pollution due to the grounding or collision of the ship, thus limiting the potential outflow of oil in case of breaches to the hull.

The 1971 amendments to the 1954/62 convention limited the hypothetical oil outflow of cargo tanks to 30,000 cubic metres anywhere in a tanker less than 400,000 dwt, with increasingly larger margins for tankers above that size.[256] The basic idea was that tankers ordered after 1972 should have cargo tanks constructed and arranged to protect against pollution after side or bottom damage. An innovative provision was made to apply the tank size limitations to all vessels ordered a year after the amendments, irregardless of whether the latter had entered into force or not, and applied to all vessels constructed after 1977. By 1973, it was found that the maritime industry had fallen into line with the 1971 IMCO amendments, and all new tankers ordered or built after 1 January 1972 had limited their tank sizes.

The latest marine pollution convention, the 1973 International Convention on Marine Pollution, as amended by the 1978 Protocol (1973/78 MARPOL convention) incorporated the 1971 amendments into the comprehensive new treaty, and also provided for damage and stability regulations.[257] Secondly, the protective location of cargo tanks used to carry oily ballast water has similarly been incorporated in the 1973/78 MARPOL convention as an additional preventive measure against accidental pollution involving tankers in ballast.[258]

Amendments to the Safety of Life at Sea Convention

The International Safety of Life at Sea (SOLAS) convention is one of the oldest and most important maritime conventions aimed at protecting human life and safety at sea. It contains provisions on the construction of ships, fire-fighting devices and procedures, life-saving equipment, communications equipment, navigational aids, survey and certification. The original SOLAS convention in 1929 has been revised to suit modern requirements by international conferences in 1948, 1960 and 1974, including a special Protocol to complement the convention in 1978.

The IMCO Assembly in 1968 approved amendments to the SOLAS convention stipulating that additional equipment for large ships was to be made compulsory.[259] The required navigational aids included gyro compasses, echo-sounders, radar, and radio equipment for direction finders.

Following the explosions on large tankers in the 1970s, the 1978 Protocol to the SOLAS convention required the use of inert gas systems for all tankers of 20,000 dwt tons or more, and required the improvement of steering systems and procedural checks onboard ships.[260]

International Convention on Loadlines

One of the 'multiplier' effects of tanker disasters resulted in the entry into force of the International Convention on Loadlines of 1966. Overloading has often been the cause of maritime casulaties, especially for cargo ships. In 1930, and again in 1966, international agreement was reached to lay down minimum freeboard, or maximum draught up to which a ship is permitted to be loaded. The 1966 Loadlines convention entered into force in 1968, and some 99 states have become parties to it by 1984.

Unfortunately, no concessions have been made in case of tankers retaining oil slops onboard under this convention. Nor have there been any requests or concessions granted to exclude space taken up by oily water separators or slop tanks in case of tankers, from tonnage measurement calculations.

Collision Regulations and Traffic Routes at Sea

In 1972 the Convention on the International Regulations for Preventing Collisions at Sea (COLREG) was revised under the aegis of IMCO. The COLREG convention sets out basic rules concerning passage of vessels at sea in order to prevent collisions. It deals with such matters as steering and sailing rules, lights and signals, sound signals and conduct in limited visibility.

Furthermore, the 1972 revision took special account of the size and limited maneuverability of tankers. Mandatory traffic separation and routeing schemes have been put in operation for some 100 areas throughout the world.

Training of Mariners

The quality of the training and attitudes of shipmasters, officers and crew ultimately determine the practical effectiveness of control measures, so that an international programme to improve the education, qualifications and selection of mariners formed part of the advances made after the *Torrey Canyon* and other large tanker disasters. It has generally been agreed that about 85% of maritime accidents are caused by human error.

In co-operation with the International Labour Organization, IMCO established a joint committee to examine mariner training programmes. In 1978 this work culminated in the Convention on Standards of Training, Certification and Watchkeeping for Seafarers (SCTW). The SCTW convention laid down basic rules for keeping navigation and engine room watches, minimum requirements for the certification of masters, chief mates and other officers in charge of communications. Mandatory rules were also stipulated for ratings and for minimum educational attainment for certification of crew.

The Right of Intervention by Coastal States

Under the then customary international law, the British government's bombing of the *Torrey Canyon* wreck could technically have been interpreted as an act of war, since they had destroyed the property of another state, a ship flying the Liberian flag, outside of British territorial waters. Fortunately, after the disaster, most states agreed that the time had come to change international law in such a way as to grant coastal states the right to protect themselves against a maritime casualty threatening or causing pollution damage to coastal areas.

In 1969 an international conference in Brussels adopted the first International Convention on Intervention on the High Seas in Case of Oil Pollution Casualties, otherwise known as the 1969 Brussels Public Law Convention or the 1969 Intervention Convention. This convention was intended to avoid the helpless plight of coastal states in case of an oil pollution incident outside their usual area of jurisdiction at sea, to define the nature of the action to be taken, and to install safeguards against any possible abuse of the right of intervention.

The basic right of intervention was provided in Article I which stated:

> Parties to the . . . Convention may take such measures on the high seas as may be necessary to prevent, mitigate, or eliminate grave and imminent danger to their coastline or related interest from pollution or threat of pollution of the sea by oil, following upon a maritime casualty or acts related to such a casualty which may reasonably be expected to result in major harmful consequences.

A second and equally important aspect of the treaty defined the obligation of the coastal state taking action before, during, and after the incident. Article VI of the intervention convention stated that the coastal state must notify the interested maritime parties, the measures taken must be proportionate to the actual or threatened damage, and finally, the coastal state was obliged to compensate the shipping interests in case the emergency measures exceeded those which were reasonably necessary at the time of the accident.

Warships and non-commercial ships owned or operated by a state were excluded from the treaty. In cases of extreme urgency, the coastal state party may act without prior notification or consultation. Finally, provision was made for the settlement of disputes through conciliation and arbitration.

In 1973 a Protocol was appended to the Convention expanding the right of intervention to cover pollution from certain substances other than oil.

The 1969 Intervention Convention entered into force in 1975, and at present some 42 states are parties to it. The

1973 Protocol took effect in 1983, and some 15 states have accepted it.

Liability and Compensation for Oil Pollution Damage

The second Convention adopted by the 1969 Brussels conference dealt with a similarly pioneering concept – that of establishing financial liability and compensation for oil pollution damage caused by maritime incidents. The 1969 Intervention Convention on Civil Liability for Oil Pollution Damage (also known as the 1969 Private Law Convention, or the Civil Liability Convention) provided for the liability of the owner of a ship which causes any damage by oil pollution, except for casualties due to war, natural phenomenon, or the negligence or act of a third party. If the incident was not the fault of the shipowner, he may limit his liability for the incident to 2,000 Poincare francs (about $140) per ton but not exceeding 210,000 Poincare francs ($14 million). However, if the incident was due to the owner's fault or privity, then he shall not be able to limit his liability.

Under the 1976 Protocol to the 1969 Convention, the unit of account was changed to the Special Drawing Rights (SDRs) of the International Monetary Fund (1 SDR was worth $1.17 at the time of the adoption of the Protocol).

In order to promote better standards of operating and maintaining ships, the 1969 Civil Liability Convention also required shipowners to maintain insurance or other financial security, and to carry a certificate onboard the vessel confirming the existence of such insurance.

The Convention applied liability to damage caused in the territory of a state-party, including its territorial sea, even when the party liable is a non-contracting state or its national. It did not apply to damage caused beyond the territorial sea, except for preventive measures intended to minimise oil pollution damage.

Litigation could be brought to the courts of the state or states in which oil pollution damage had occurred, and an owner wishing to limit his liability must establish a fund with one such court which will have the competence to determine the issues on compensation.

Recognizing that potential victims of oil pollution were not yet fully protected, a supplementary agreement was adopted later – the 1971 International Convention on the Establishment of an International Fund for Compensation for Oil Pollution Damage. Under the 1971 Fund Convention, states or persons who suffer oil pollution damage would be compensated if they were unable to obtain relief from the shipowner, or if the latter's compensation was not sufficient to cover the damages suffered.

The fund has been financed by levies on persons or companies who import or receive oil in contracting staes. Again, shipowners must carry a certificate attesting to the availability of insurance or other financial security. The amount compensable per incident was limited so that the aggregate amount paid under the 1969 Convention and the 1971 Fund should not exceed 450 million Poincare francs ($35 million), although the Fund Assembly can raise that figure to a maximum of 900 million gold francs ($70 million). Under the 1976 Protocol to the Fund, these figures were converted to 30 million and 60 million SDRs respectively.

In 1978 the Fund came into official existence and to date some 30 states have become parties to it. Various claims have successfully been settled for oil pollution incidents since 1979.

Further revisions to the 1969 and 1971 compensation conventions were made by an international conference held in London on 30 April to 25 May 1984. These changes became necessary due to advances in the international law of the sea and the experience with more expensive pollution accidents like the *Amoco Cadiz.* Although the 1984 conference failed to agree on a treaty to cover liability for seaborne hazardous and noxious substances other than oil, the delegates from 60 governments nonetheless approved two Protocols amending the 1969 Civil Liability and 1971 Fund Conventions respectively.

Thus, the 1984 Protocol to the Civil Liability Convention included the following major amendments: (1) Minimum liability limits were established at 3 million SDRs for small tankers of 5,000 gross tons, rising to 420 SDRs per ton up to a maximum of 59.7 million SDRs for larger tankers (Article 6); and (2) the geographical scope of coverage may now include the Exclusive Economic Zone or comparable area under international law, and not just the territorial seas of state-parties (Art. 3).

Other significant changes affected the definition of ship (Art. 2(1)), the definition of oil (Art. 2 (2)), the meaning of pollution 'incident' (Art. 2(3)), and the concept of pollution damage (Art. 2(3)). A blanket of immunity was extended to third parties (Art. 4(4)), and insurance certificates were to be revised accordingly (Art. 7).

The 1984 Protocol will come into force 12 months after ten states have ratified it, including six states with one million or more gross tanker tonnage each. Afterwards, the Legal Committee of the International Maritime Organization may propose changes to update the liability limits.

The 1984 Fund Protocol features the same revised definitions found in the 1984 CLC Protocol, as well as the same geographical scope and unit of account. The indemnification function of the 1971 Fund Convention was deleted, and the shipowner's liability would now roughly equal that of the cargo owner. The Fund will still compensate for damages when no liability arises under the CLC, or when the damage exceeds the CLC limit, or the shipowner is insolvent. The coverage for compensation will rise to 135 million SDRs when the 1984 Protocol enters into force with eight or more states contributing the value of 600 million tons of imported oil. At such time as any three state-parties together import 600 million tons of 'contributing oil,' the ceiling will be raised to 200 million SDRs expanded coverage (Art. 6(3)).

A transition period coverage is still ensured for claimants, anticipating treaty law problems of co-ordinating the entry-into-force of the two separate Protocols. Meanwhile, the Fund's Executive Committee will be dissolved, leaving the Assembly (with state-parties as members) and the Secretariat to run the organization. At present, some 29 states are parties to the original Fund Convention.

Voluntary Compensation Schemes

Prior to the adoption of governmental arrangements for oil pollution liability and compensation, tanker owners and oil companies initiated separate and voluntary compensation measures for victims of oil pollution damage. These private remedies by the industry were mainly intended to stave off further criticism by the public or the possibility of stricter unilateral measures by governments. Presently, they also serve as alternative modes of securing compensation for oil pollution victims who do not wish to enter into court litigation or diplomatic representations, as they would have to do through the 1969 and 1971 Conventions.

Under the Tanker Owners Voluntary Agreement Concerning Liability for Oil Pollution (TOVALOP), some 90% of tanker owners have agreed to clean up oil pollution themselves or to reimburse governments and public agencies for any oil pollution damage. Liability is limited to about $100 per registered ton of the vessel or around $14 million per vessel per incident.

A second private fund was set up by the oil industry under the Contract Regarding an Interim Supplement to Tanker Liability for Oil Pollution (CRISTAL). Some 80% of the international oil industry (refineries, local oil companies, etc.) have provided additional funds for the extension of liability for each incident up to $30 million worth of damages.

In 1974 alone, TOVALOP settled over two hundred claims for private compensation by oil pollution victims, and by the same year CRISTAL had settled some 12 out of 65 claims, averaging $400,000 in all.[261]

These private industry initiatives, and the inter-governmental agreements on liability and compensation, represent contemporary endorsement of the 'polluter pays' principle.

Mitigation and Clean-up Measures

The experience with large-scale oil pollution in the past two decades has resulted in the promotion and development of various means of coping during emergencies in an attempt to reduce or mitigate the damage to coastal resources, amenities and wildlife.[262] In this new growth industry of cleaning oil spills and marketing suitable pollution control devices, equipment and methods, ironically the oil industry itself has played the paramount role. Many leading countries like Britain, the United States, Canada, and others have developed special national contingency plans and anti-pollution units to deal with oil pollution emergencies.

Generally, the information on mitigation measures involves knowledge of the types and behaviour of oil spilled at sea, and the containment, removal and treatment of oil spills at sea or on shore by chemical, mechanical or manual methods. Technical information on these matters are provided by numerous private companies, oil industry groups, leading governments and international organisations, especially IMO which publishes a special manual on the subject.[263] Various conferences and workshops have also regularly discussed current procedures and problems.[264]

Regional Arrangements

Since 1969 regional arrangements have been developed for areas vulnerable to pollution, e.g. the 1969 Bonn Agreement in Dealing with Oil Pollution of the North Sea by Oil, the 1977 Baltic agreement, etc.[265] These arrangements have included co-ordination and information-sharing on prevention and safety measures, contingency plans, mitigation methods, and general environmental policy. The scope of regional arrangements has been broadened to cover not only oil pollution but other types of marine pollution, as well as pollution from land-based sources.

The United Nations Environment Programme maintains a Regional Seas project to develop special regimes for the seas and oceans, and 'Save the Seas' action programmes have been started for the Mediterranean, the Gulf, West Africa, the Caribbean, Red Sea, South-east Asia, South-east Pacific, South-west Pacific, East Africa, and South-west (Latin American) Pacific.

Under the 1973 MARPOL convention, five special areas – the Mediterranean, Black Sea, Baltic, Red Sea and the Gulf – have been designated as 'special areas' in which only clean operational discharges are allowed from ships. A 'special area' was defined to mean a sea area where for recognised technical reasons (its oceanographical and ecological condition and strategic traffic location) the adoption of higher standards for the prevention of sea pollution by oil was required.[266]

It is within the context of these regional arrangements that problems peculiar to the area can presumably be more easily resolved than in the wider context of international agreements.

7

THE 1973 MARINE POLLUTION CONVENTION

Our study of oil pollution control ends with the current treaty on the problem – the 1973 International Convention on Marine Pollution and its 1978 Protocol. We shall examine some of the major issues at the 1973 London Conference and then analyse the Convention and Protocol. Overall, the 1973 Convention and 1978 Protocol represent what most experts today agree to be the best possible solution to meet the problem of marine pollution. Whilst some may argue that the measures yet fall short of the desired goal of having appropriate conservation techniques to accompany technological progress, most others believe that full implementation of the legal framework would result in the satisfactory prevention of pollution by ships. Yet there are some serious flaws in the 1973 Convention and 1978 Protocol that, even with implementation of their provisions, the legal framework may not meet the very purpose for which it was intended.

The 1973 London Conference on Marine Pollution

Over 600 delegates from 71 countries met at London's Church House from 8 October to 2 November 1973 to attend the International Conference on Marine Pollution sponsored by IMCO. In view of the growing concern for marine pollution, not only by oil but other noxious or hazardous substances and the urgency of resolving technical approaches to the problem, the 1973 diplomatic discussions acquired great importance.

Among the high objectives of the Conference were (1) to draft a comprehensive new convention that would completely eliminate the wilful and intentional discharge into the seas by ships and other marine craft of oil and noxious or hazardous substances other than oil, and the minimisation of accidental spills by all types of ships at sea; (2) to achieve by 1975 if possible, but certainly by the end of the decade, the complete elimination of pollution by normal operations of ships; and (3) to expand the 1969 Brussels Intervention Convention to cover other types of substances causing pollution incidents at sea.

The growing influence of the coastal and Third World countries was reflected in the list of officers chosen by the Conference, with Mr. S.V. Bhave, Director General of the Indian Shipping Ministry, as president. For the first time, the Third World delegations outnumbered those of the more-developed nations at an oil pollution conference. The trend towards accommodating the interests of developing states at maritime fora was confirmed by the appointment of Mr. C.P. Srivastava, president of the Indian national shipping company, as IMCO Secretary-General after the retirement of Sir Colin Goad of Britain in 1974.

To facilitate the work of the Conference, four working committees were established, apart from the drafting and credentials committees. Committee I considered the drafting of the main Articles of the Convention and other legal matters referred to it. Committee II dealt with the technical regulations on oil pollution, sewage and garbage. Committee III dealt with chemicals and other harmful substances in package form or containers. And Committees IV drafted the Intervention Protocol to the 1969 Brussels Convention.

At an early stage, the Conference decided to draft an entirely new treaty which would supercede the 1954 Convention and its amendments, but incorporated many of the amendments in 1969 and 1971. The following sources of marine pollution were specifically excluded: (1) dumping licenced by a state under the terms of the 1972 London Dumping Convention; (2) pollution arising from the exploration, exploitation and associated off-shore processing of seabed and coastal mineral resources, which would be covered by national and regional regulations; and (3) pollution arising from scientific experiments or clean-up operations. The technical scope of the new regulations nonetheless seemed broad. All merchant ships from hydrofoil and air-cushioned vehicles to tankers and other ocean-going liners, as well as submersibles, fixed and floating platforms, fell under the regulations.[267]

There were two major issues confronting the delegates – firstly, to define the means of enforcing the Convention in light of the growing demand for greater powers to be granted coastal authorities at the expense of the traditional control of flag states for their vessels, and secondly, to make the technical regulations for controlling marine pollution more effective and comprehensive. Overall, it may be said that the 1973 Conference made a brave attempt at the first hurdle, but it reached such an unsatisfactory compromise at the second that another technical conference in 1978 met to sort out the problems left unresolved by the technical provisions in the new Convention.

Due to the forthcoming United Nations Law of the Sea Conference in Caracas, Venezuela, the marine pollution Conference delegates were under instructions not to prejudice their country's position at the Caracas talks. Thus, the Article on enforcement (Article 4) was drawn up in such a way as to leave the interpretation of the term 'waters within its jurisdiction' of the coastal state vague enough to include any extension of territorial waters at the UNCLOS negotiations. As part of a 'package deal' involving other Articles, Committee I voted unanimously for a proposal that empowered any state-party to penalise a foreign ship polluting the 'waters within its jurisdiction,' subject only to an obligation to inform the flag state of the details of the case.[268] But the flag state retained the option to prosecute if and when it is asked to do so against one of its ships.

However, the 'package deal' began to unravel when the Canadian delegation introduced their proposal for universal port state jurisdiction which would have enabled even a third state to act against a foreign vessel accused of polluting the waters of a state-party. The Canadian proposal was sufficiently diluted to make the innovation more palatable to the delegates with large shipping interests, e.g. only monetary fines would have been imposed, the ship would be released after posting a bond, etc.[269] But due to the hardcore opposition of important maritime powers and the Socialist bloc, the very idea of port state jurisdiction proved an anathema to the majority.[270] As one European shipping delegate stated, 'I will never accept that pollution is as serious a crime as piracy or slavery.' The 1962 Conference opposition to port state authority, as was seen in Part Five, also nearly broke up discussions on legal jurisdiction.

The London Conference was also expected to clarify the extent to which states may take special measures in advance or in excess of international standards in order to protect themselves. It will be recalled that the original 1954 Convention contained the following clause on the residual rights of states:[271]

> Nothing in the present Convention shall be construed as derogating from the powers of any Contracting Government to take measures within its jurisdiction in respect of any matter to which the Convention relates or as extending the jurisdiction of any Contracting Government.

Thus, the original Convention acknowledged the powers of states to legislate only within their own competence for their own vessels or within their territorial waters, but expressly prohibited regulations which feel under the jurisdiction of another state or which exceeded the international standards under the treaty.[272]

Such was the situation when, after the *Torrey Canyon* and other tanker casualties causing oil pollution disasters, a number of states like Britain, Canada and the United States took matters in their own hands and passed legislation which, if they did not actually exceed the standards prescribed in international treaties, at least empowered these states to take measures in advance of available international standards. Evidently, such 'leap-frogging' legislation forced the issue of reconciling unilateral action with the more cumbersome method of reaching internationally-agreed regulations.

The proposals at the 1973 Conference generally fell into two categories: (1) those wanting to give states greater freedom of action on such matters as discharge standards, manning, equipment, ship design, etc., and (2) those seeking to restrict the powers of staes to within internationally-agreed criteria.

The search for a consensus to resolve this issue nearly caused the Conference to end in a deadlock. The proposed new article on special measures sponsored by Canada and thirteen other delegations was passed at Committee stage despite the strong American objections.[273] Behind the scenes, the American delegation let it be known that if the Conference passed the proposed new Article, they would find it difficult to sponsor ratification of the new treaty by the U.S. Congress. The Canadians however organised their own diplomatic demarche to sustain interest in the draft Article, by which the Canadians wished to legitimise their special claim to protect the Arctic environment.

As the discussions continued at Plenary Session stage, it seemed uncertain whether any one opinion would prevail. Since a two-thirds majority was needed to pass the new Article, those wishing to limit the residual powers of states, and thus delete the Article altogether, needed only a blocking third of the votes. The final vote at Plenary fell short of the required two-thirds majority, thus leaving it up to the Law of the Sea Conference to resolve the issue.[274]

Apart from the juridical issues, the Conference also discussed the technical regulations to control marine pollution by ships, which were placed into the Annexes appended to the Convention. By separating the technical provisions into Annexes, it was the desire of the delegates to facilitate the implementation of the Convention and any subsequent amendments. The most controversial and lengthiest deliberations on the technical Annexes were held at Committee II. where those wanting to improve upon the 1969 and 1971 changes to the original oil pollution treaty introduced additional proposals such as the inclusion of white (non-crude and lighter grade) oils, segregated ballast, double bottoms, and monitoring systems into the scope of the new Convention. By contrast, the technical Annexes on chemical pollution sewage and garbage – which were included for the first time in international control – were mostly based on the regulations previously drafted by IMCO working committees preparatory to the Conference. The content of these new regulations will be discussed later.

Due to the technical complexities of the new Convention, it is just as well that the Conference agreed to pass a new Article (Article 17) on the promotion of technical co-operation. Although initially regarded with suspicion by the developed countries as the less-developed states' way of asking for a hand-out in future, the proposal in fact anticipated the need for increased activities in sharing technical expertise and research on a scale never before undertaken. As anticipated by its sponsors (including the present writer who initiated the idea), the new Article on technical co-operation has equally served the interests of the more-developed countries and the International Maritime Organization and United Nations Environment Program (see Fig. 5).

On looking back, the 1973 Conference produced the world's first 'comprehensive' treaty on marine pollution, with technical regulations covering not only oil pollution but other marine pollution problems as well. As such, this was an advance on previous sea pollution treaties limited to oil. However, despite the veneer of accomplishment shown by the new Convention on marine pollution, events would show that the Conference had failed to meet its major objectives. It did not produce that 'comprehensive new Convention' to completely eliminate operational pollution; nor did the new treaty meet the target date of international control that had previously been set by oil pollution treaties. The irony is that these international standards required major alterations some five years later before the new Convention had been ratified.

Figure 5. Technical Co-operation in Action

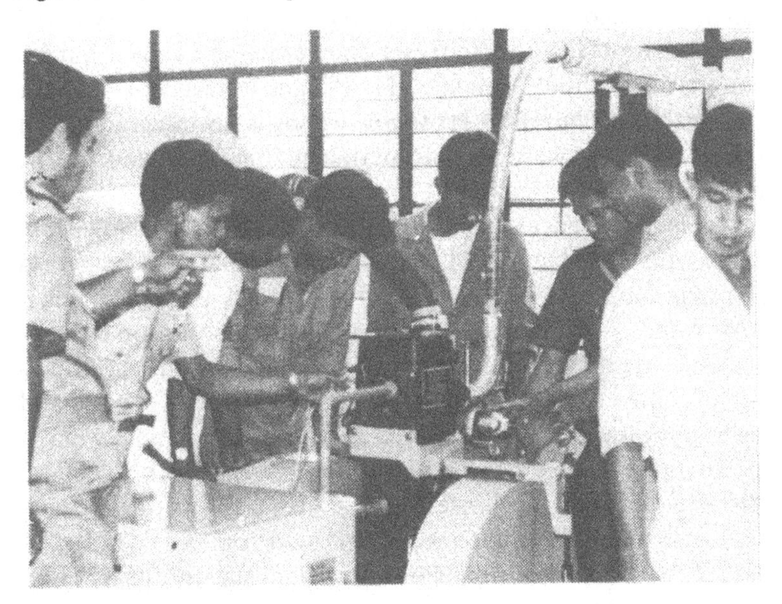

Source: IMO News No. 4:1984

Thus. in 1978 a special Tanker Safety and Pollution Prevention Conference was held again in London to draft amendments to the oil pollution annex of the 1973 Convention. The Protocol arising from this special conference became an integral part of the 1973 Convention, and acceptance of the 1973 treaty by states would have to include acceptance of the 1978 Protocol as well.

The 1973/78 Marine Pollution Convention

The 1973/78 Marine Pollution Convention is the world's first treaty to regulate all forms of marine pollution by ships, with the exception of the licensed dumping at sea of land-generated wastes under the 1972 London Dumping Convention. The 1973 Convention itself consists of twenty Articles, and two Protocols dealing with the obligation to report incidents and with arbitration. and five Annexes – those containing the regulations for oil pollution (Annex I): chemical pollution (Annex II); pollution by harmful substances carried in packages portable tanks, freight containers, or road and rail tank wagons, etc (Annex III); sewage (Annex IV); and garbage (Annex V). Annexes I and II of the Convention are mandatory, while Annexes III, IV and V are optional for state-parties. The 1978 Protocol, which is an integral part of the 1973 Convention, strengthened and expanded the requirements of the parent Convention.

The Convention came into force on 2 October 1983, and as of 1985, some 31 states representing 72% of world shipping have ratified the main Articles and the two mandatory Annexes. The regulations on chemical pollution (Annex II) will enter into force on October 1986. A significant number of states have enacted national legislation on sewage and garbage pollution by ships, even though the technical Annexes on these problems remain optional. Some twenty countries have ratified optional Annexes III and V of the Convention, and nineteen countries have ratified Annex IV.

The 1973/78 Marine Pollution Convention differs from the 1954 treaty in the following ways: the regulations against marine pollution include substances other than oil; the definition of oil pollution covers all petroleum oil except petrochemicals; the discharge standards for new tankers have been reduced by half; no discharges are permitted inside 'special areas' in the Baltic, Mediterranean, the Black and Red Seas, and the Arabian/Persian Gulf; authorised discharges from non-tankers may be made only at a distance of more than 12 miles from shore; all tankers of more than 150 gross tons are required to have monitors checking the details of the rate and amount of discharge; in new tankers the monitoring systems should have automatic shut-off devices; a more detailed Oil Record Book; all new tankers of 70,000 dwt tons or above must have segregated ballast tanks; the provision of adequate reception facilities (see Fig. 6) extends to oil loading terminals, repair ports and other ports where ships have oily residues to discharge; a limited form of port state jurisdiction is allowed for the inspection of ships, but the flag state retains legal and administrative sanctions on its vessels, except for offences committed within the coastal state's jurisdiction.

Figure 6. Port Reception Facilities

Source: IMO News No. 1:1984

Our analysis of the 1973/78 Convention will include many of its recognised merits in controlling marine pollution. but we shall also discuss certain problems that deserve attention and which may have been overlooked before.

Legal Aspects of the 1973/78 Convention

The effectiveness of an oil pollution treaty depends

upon two important factors – firstly, its acceptance and implementation by governments, and secondly, the co-operation of the oil and shipping industry in complying with the regulations. Whilst there are other factors predisposing government officials and industry leaders in favour of pollution control measures (which shall be discussed in the next section), ultimately it is the authorities and the so-called 'culprits' who must integrate and practice the values of conservation.

Due to the experience with previous anti-pollution regulations, the present trend has been to tighten up on the loopholes whereby violators have acted with impugnity in the past. It is encouraging to note that the 1973/78 Convention, in conjunction with the 1982 Law of the Sea Convention, contains various provisions to ensure stricter enforcement of the regulations. However, the legal interpretation of the 1973/78 Convention, which has entered into force, hinges in some ways upon the 1982 Law of the Sea Convention, which has not yet come into operation. Such an anomaly becomes even more pronounced when states enact anti-pollution standards for wide areas off their coasts.

To ensure that the 1973/78 Marine Pollution Convention is enforced, Articles 4 to 6 provide for three types of jurisdiction. Firstly, the flag state of the vessel may prosecute for offences wherever they occur and shall ensure that their ships comply with the treaty's technical requirements. Secondly, the coastal state may prosecute for discharges in 'waters within its jurisdiction'. The exact limit of this competence 'shall be construed in the light of international law in force at the time of the application or interpretation of the Convention' (Article 9). Thirdly, a port state may inspect a foreign vessel's certificate, or the condition of the vessel itself if so warranted, but it may not prosecute the vessel for contraventions it has discovered as a result of the inspection. The port state may only inform the flag state of the vessel's deficiencies. Furthermore, Article 7 provides that compensation be given for any undue delay to a vessel as a result of the inspection.

The International Maritime Organization has developed various guidelines on the manner of reporting illegal discharges to flag states, and the Convention itself contains Protocol I, which specifies the obligation and method of reports on incidents involving pollution.[275] Although third party or port state inspection of vessels does not include the authority to prosecute foreign vessels, this innovation offers an additional preventive measure when it is applied by parties to the Convention. On 26 January 1982, the nine member states of the European Community signed a Memorandum of Undertaking on Port State Control for the co-ordinated system of inspecting foreign vessels visiting their ports, as well as that of Spain, Potugal, Finland, Norway and Sweden. Out of some 9,847 inspections conducted on 7,350 ships of 106 flag states, of about 20% of the world fleet, it was found that 115 (or 0.85%) had deficiencies in marine pollution standards, mostly on the Oil Record Book and some on oily water separators and retention on board procedures.[276] Evidently, other regional arrangements of a similar nature may be adopted elsewhere.

The nature of coastal state jurisdiction, however, remains unclear. Under the 1982 Law of the Sea Convention, whch was signed by 118 nations in Jamaica but which has not yet entered into force, the definition of 'waters within its jurisdiction', or the territorial sea, of a coastal state would be measured as 12 nautical miles from coastal baselines rather than the traditional 3 nautical miles from the coast.[277] In addition, the 1982 LOS Treaty authorises coastal states to apply anti-pollution regulations within their Exclusive Economic Zone (EEZ) a sea area which may extend up to 200 nautical miles from the territorial sea under certain conditions.[278] Special measures for ecologically-sensitive areas are authorised only for 'ice-covered areas', an accommodation of the Canadian concern for the Arctic.[279]

In cases of dispute on coastal state jurisdiction over foreign vessels in their EEZ, international lawyers can only express the opinion that the 1982 LOS Treaty should be regarded as having acquired the status of customary law, if not conventional law. Opposition by the United States, Britain and other important maritime states to the seabed provisions of the 1982 LOS Treaty has prevented its entry into force. Meanwhile, states continue to designate EEZs off their coast by means of national legislation, and it seems certain that coastal states will apply the 1973/78 Marine Pollution Convention protection to their EEZ apart from their territorial sea.

To forestall potential legal disputes about the geographical scope of the 1973/78 Marine Pollution Convention, it would be better to specify the inclusion of Exclusive Economic Zones within the ambit of coastal state jurisdiction. A specific amendment to this effect has already been provided under the 1984 Protocol to the 1969 Brussels Intervention Convention.[280]

The ratification and enforcement of a pollution treaty by governments may only be as effective as the willingness of the oil and shipping industry to implement its provisions. In this regard, oil pollution control has benefited from two trends in the past decade – the slump in the tanker industry and the high cost of oil. In the first place, the industry was forced to phase out many tankers which would not meet the 1973/78 Convention standards. Secondly, the high cost of oil made conservation techniques more attractive and feasible than before.

Hence, despite previous disappointing compliance by the industry of oil pollution regulations, major operators have largely accepted various voluntary codes contributing to the general reduction of oil pollution, which include the 'Clean Seas Guide' and 'Monitoring Load on Top.'[281] In some countries like the Soviet Union, Greece, Italy, Singapore and Bahrain, laid-up tankers have been used for dirty ballast and slop oil reception, a practice which the industry commends elsewhere and, as we have seen in Part One, is not an entirely new idea. Meanwhile, industry spokesmen continue to remind governments of their obligation to provide reception facilities for waste oils and to warn of serious legal disputes developing between parties to the Convention because of the failure of some governments to compel their own interests to comply with their part in the international agreement.[282]

Technical Aspects of the 1973/78 Convention

The irony of pollution treaties has been shown by the fact that departures from the principle of preventing pollution at the source have actually been written into these agreements. As Fig. 7 shows, the technical means of preventing oil pollution have ranged from the least effective designation of oil-free zones beyond which discharges are authorised to the ideal of retention on board techniques through port and tanker modifications. The early attempts to curb pollution by ships only revealed the inadequacy of the zone system, and contemporary control measures now incorporate variations of the more effective technical methods. Whilst the 1973/78 Convention requires retention on board techniques, it yet authorises pollution at sea, albeit under limited conditions and outside of special areas. Furthermore, it has applied certain principles of the Load on Top method for oil pollution control – a highly dubious practice – to the control of chemical pollution at sea, thus extending a dangerous and irresponsible precedent.

Annex I (Oil Pollution) of the Convention maintains substantially similar discharge criteria to those specified in the original oil pollution convention of 1954, as amended in 1969 However, certain provisions have strengthened the oil pollution requirements. These are summarised, as follows: (1) the definition of oil includes petroleum in any form including crude oil, fuel oil, sludge, oil refuse and refined oil products other than petrochemicals; (2) for new tankers, the total quantity of oil which may be discharged into the sea has been halved to 1/30,000 of the total quantity of the cargo carried in the previous voyage; (3) when discharging oil, tankers and other ships must have in operation an oil discharge monitoring and control system and oily-water separator or filter equipment; (4) within special areas, including the Mediterranean, the Baltic and Black Seas, the Red Sea, and the Gulf area, only segregated or clean ballast may be discharged; and finally (5) parties to the Convention are obliged to ensure the provision of adequate reception facilities for residues and oily mixtures at oil loading terminals, repair ports

Figure 7. Solutions to Tanker Operational Pollution

A. Prohibited Zones

To prohibit discharges deemed to be harmful within fixed outward distances from shore but to allow uncontrolled discharges outside prohibited areas.

Problems:

1. Political vagaries of fixing zones.
2. Enforcement of zones.
3. Not advisable for highly toxic and bioaccumulative pollution.
4. May only defer the ultimate fate and effects of pollution.
5. Aggregates discharges in major shipping lanes, usually close to major trading countries with more articulate and environmentally-conscious communities.
6. Not advisable for enclosed seas where coastal pollution contributes to marine pollution problems.
7. A known palliative.

Result: Transfer of pollution/maximum loss of cargo dregs

B. Load on Top (LOT)

To eliminate the bulk of seawater, consolidate undeballasted wastes with tank washings, to control rates and quantities of discharge, and retain most cargo dregs, either for:

B-1.	Recycling wastes onboard.	B-2.	Discharge at loading port (LOT + partial retention on board until disposal at next port.)	B-3.	Dedicated loading (LOT + dedicating tanker to load similar or compatible cargoes.)
B-1a.	LOT + oil as fuel.				
B-1b.	LOT + tank cleaning with oily slops.				

Problems:

1. Legal enforcement of permissible rate and quantity of discharge, otherwise the need to involve the crew in monitoring and controlling the discharge and retention operations.

B-1a.	2. Not suitable for non-fuels. 3. Accumulated sludge.	2 to 8	All the problems associated with (D), although reduced to	2 to 4.	All the problems associated with (C-2), but, in addition,

Figure 7. Solutions to Tanker Operational Pollution (cont'd)

B.1b. 2 to 3. Hazard and tanker space problems associated with (C-1 nos. 1 to 7)

4. Only suitable if previous cargo can be used accordingly.

Result: Reduced pollution/reduced losses for LOT methods.

the extent that the tanker has previously partially deballasted and discharged some of its cleaning water.

the added quantities of ballast wastes to be discharged at sea.

C. Segregated ballast combined with other methods

To rationalize vessel design and operations to the extent that ballast water is segregated from cargo residues, in combination with:

C-1. Closed cleaning cycle (SBT plus retention of tank washings onboard or Crude oil washing.)

Problems:

1 to 7. All the problems associated with (D), although reduced to the extent that ballast dregs would not normally be included in the quantities of ships' wastes to be kept onboard and disposed ashore.

8. Extra care during tank cleaning, and properly inerting tanks with combustible gases.

Result: No pollution/no losses

C-2. Dedicated Tanker

SBT plus obviating tanker cleaning and dedicating tanker to load similar or compatible cargos.

Problems:

1. Inadvisable for toxic or incompatible mixtures.
2. To what extent can the bulk liquid trade coordinate and rationalize tanker operations accordingly?
3. Build-up of sludge can add to the occasion and fees for drydock maintenance.

Result: No pollution/no losses

C-3. Load on Top (LOT)

SBT plus controlled overboard discharge of tank washings.

Problems:

Depending on what the fate of the retained mixtures are, all the problems associated with the types and sub-types of (B-1) and (B-2), but reduced to the extent that ballast wastes can normally be disregarded.

Result: Reduced Pollution/reduced losses.

D. Total cleaning at port of discharge

To clean the tanker at the port of discharge before it takes in ballast and loads another cargo.

Problems:

1. Are the facilities available and adequate?
2. Are coastal authorities prepared to assume responsibility for what can be a hazardous and time/space-consuming tank?
3. Do port incentives exist for tankers to avail of local plants and services at little or no cost to operators?
4. Are ship operators prepared to lay off their tankers for the time it takes to clean vessels?
5. Are the recovered wastes useful, or are they likely to pose a secondary pollution problem?

Result: No pollution/no losses

E. Retention on Board (ROB) or total cleaning at loading port.

To sail, ballast and clean tanks as usual, but to retain all dirty ballast and washings onboard until their total disposal at the next loading port.

Problems:

1 to 5. All the coastal and shipping problems associated with (D).

6. In addition, the hazard of sustaining a highly charged atmosphere during the ballast journey if the wastes and mixtures are combustible or toxic.
7. Tanker space for consolidated wastes.

Result: No pollution/no losses.

and in other ports in which ships have such residues to discharge.

In order to comply with the above discharge requirements, oil tankers must operate in accordance with Load on Top procedures. Under the procedure, which has been described more fully in Part Five, the tanker consolidates its wastewater contaminated by oily residues and partially decants them onboard. After the separation of the oily dregs in the mixture, the tanker proceeds to discharge the free water in the mixture and to retain most of the dregs in the slop tank. However, because some oil yet becomes entrained in the dirty ballast and in the effluent from the slop tank, discharges are permitted only beyond 50 miles from land outside of special areas.

In addition, the 1973/78 Convention introduces certain requirements for the construction and equipment of ships in order to prevent operational discharges of oil and to the control oil pollution due to tanker accidents. These new requirements are briefly summarised, as follows: (1) oil tankers must be fitted with oil discharge and monitoring equipment, with a recording device to provide a continuous record of the discharge; (2) any ship of 400 gross tons or more must be fitted with an oily-water separating equipment or filtering system; (3) oil tankers must be fitted with a suitable slop tank with the capacity to retain the slops from tank washing and dirty ballast; (4) new oil tankers of 70,000 deadweight tons or more must be provided with segregated ballast tanks of suitable capacity to enable them to operate safely on ballast voyages without recourse to the use of oil tanks for water ballast except during severe weather conditions; (5) new crude oil tankers of 20,000 deadweight tons or more, and new product carriers of 30,000 deadweight tons or more, must be provided with segregated ballast tanks (SBT) which must be protectively located to prevent oil pollution during collisions or strandings at sea; (6) new crude oil tankers must be provided with crude oil washing (COW) system (whereby crude oil instead of water is used to wash residues in the tank walls) and inert gas system (IGS) to prevent gas explosion in the tanks during tank cleaning with crude oil, as specified in the 1978 Safety of Life at Sea Protocol; (7) oil crude oil tankers of 40,000 deadweight tons or more must be provided with segregated ballast tanks, especially dedicated clean ballast tanks (CBT) or crude oil washing, while old product carriers of similar tonnage must have either segregated ballast tanks or dedicated clean ballast tanks; (8) the size and arrangement of cargo oil tanks are limited to minimise the outflow of oil in cases of collision or standing; and finally (9) new oil tankers must comply with the subdivision and damage stability requirements to ensure that they can survive side or bottom damage during accidents.

Annex II of the Convention sets out detailed requirements for the discharge criteria and measures for the control of pollution by noxious liquid substances carried in bulk. For this purpose, noxious liquid substances are classified into Categories A, B, C and D depending upon their hazard to marine resources human health, amenities and other uses of the sea. Some 250 substances were evaluated for this purpose, and those falling under the four categories are included in the list appended to the Convention. The discharge criteria for noxious liquid substances are determined according to the hazardous nature of the substance, with virtually nil discharges allowed for Category A (highly toxic) substances and in some cases unlimited control for Category D (minimal harm) substances.

The Baltic and the Black Seas are designated as special areas for Annex II substances, in which stricter restrictions are applied. With regard to construction and equipment, all chemical tankers are required to comply with all the provisions of the International Bulk Chemical Code (IBC Code).

Annexes III, IV and V remain optional parts of the Convention which refer to other sources of marine pollution. Annex III contains general requirements for the prevention of pollution by harmful substances carried by sea in packaged form or in freight containers, portable tanks or road and rail tank wagons. Detailed provisions on packaging, marking and labelling documentation, stowage, quantity limitations and other aspects aimed at preventing or minimising pollution of the marine environment by such substances are developed within the framework of the International Maritime Dangerous Goods Code or in other guidelines. Under Annexes IV and V (sewage and garbage), ships will not be permitted to discharge sewage within four miles from the nearest land unless they have in operation an approved treatment plant. Between four and twelve miles from land, sewage must be comminuted and disinfected before discharge. Specific minimum distances from land have also been set for the disposal of all the main types of garbage from ships. The disposal of all plastics is prohibited, since these are not biodegradable. In special areas, stricter procedures are applied for the disposal of garbage.

Thus, it seems curious in the face of continuing reports on pollution by ships at sea that the 1973/78 Marine Pollution Convention in fact specifies further doses of pollution, albeit under certain conditions. It will be recalled that the classic defence about the capacity of the oceans to provide a vast sink for the disposal of wastes from ships has been disputed since the idea of controlling vessel-source pollution was first discussed. The justification for authorizing discharges from ships at sea has been based on the scientific contention – mostly sponsored by the oil and shipping industry – that discharge standards must be related to the 'size of the area or volume receiving the effluent,' rendering the tanker's minimal pollution in 'harmless and greatly dispersed form.'[283] Other discharge sponsors assume that the 'turbulence from propeller action in the ship's wake will atomize or divide' pollution at sea, thus enhancing the natural disposition of the substance.[284] Much of the scientific studies submitted to the 1973 Conference have accordingly been oriented along these lines, and it is significant how little these same studies actually support such justifications for continued discharges at sea.[285]

Yet this strange dogma persists, and against which circumstantial reports of chronic pollution are more readily ascribed to non-Convention ships and even natural seepages or land-based pollution but not to the discharges authorised under existing agreements. That ships may disperse soluble wastes in their wake, a principle underlying the justification for the LOT system, has subsequently been carried as a working principle on which to base the deliberate pollution by ships of other noxious liquid substances, which, unlike oil with its disputed persistence at sea, possess inherently more hazardous constituents.

Accordingly, the 1973 Convention Annex on chemical tankers requires different limits and procedures for the routine discharges of such wastes in an ascending degree of premissibility. It is only when carrying Category A (highly toxic) substances that convention tankers must retain all dirty ballast and tank washings for disposal into port facilities. Outside of the two special areas, the Baltic and the Black Sea, tanker discharges of residues of Categories B and C substances (less toxic and harmful) are allowed under certain conditions. A pre-cleaning process is required of Category B carriers within the two special areas, while Category C tankers must only ensure that the concentration of the substances in the ship's wake does not exceed 1 ppm in such areas. Category D substances (recognizable hazards or affecting minimal harm) may be discharged in all areas provided certain conditions are met.[286]

Masters would have to ensure that suitable arrangements are found to meet the conditions for discharging contaminated wastewater from dirty ballast and tank washings. In many cases this means slop tank treatment of residues encompassing about 66% of Category B and 58% of Category C products, and pre-voyage planning.[287]

Since prior to the 1973/78 Convention, no formal regulations were contemplated against operational pollution by chemical tankers, the rules represent an improvement on the efforts to balance environmental and maritime interests. After all, in its effective accommodation of the two components at issue – environmental and maritime industry problems – lies the function and in a sense the main strength of any policy to manage shipping pollution. The question remains however of whether it will be possible to achieve effective results when discharge standards suppress the essence of environmental quality *and* pose complex commands for mariners who may wish to abide by the Convention but are otherwise prevented from doing so due to lack of port facilities, or due to cost-benefit considerations, or the lack of suitable training and expertise of mariners.

Clearly, environmental policy-making calls for an historical perspective rather than sheer speculation on anticipated results from measures which may defeat the purpose for which they were intended. In many ways the 1973/78 Convention expresses recognition of the dangers and the difficulties attached to marine pollution control. For special areas and highly toxic substances, the intention is evidently to move toward the highest possible control standards. On the other hand, the Convention evokes the longstanding proclivity to defer port and tanker adaptations until the irony becomes more apparent when open areas, e.g. the North Sea, the Pacific, etc., produce evidence of serious damage and when depletable resources become too expensive to waste. Whichever the case, the story will have to be taken back to where it first began – with the radical reformation of port and tanker operations.

The North Sea as a Special Area

The North Sea, which has the unseemly distinction of losing its former right to a special area status under the 1973/78 Convention (one denoting higher standards of environmental protection) is precisely the major international waterway in the world most susceptible to vessel-borne discharges of oil and chemical wastes. Since proposals for including the North Sea as a special area under the Convention have been discussed by governments and concerned groups during the past few years and a diplomatic conference in London in 1987 is expected to settle the matter it is timely to recall that such a move has been proposed by the author as far back as 1978.[288]

The North Sea became a special area of environmental protection under the 1962 amendments to the 1954 oil pollution treaty, as previously discussed in Part Five of this book. Earlier efforts had failed, notably in 1929 and another at the 1954 London Conference.[289] The total ban on pollution in practice applied only to oily washings discharged from vessels during deballasting and tank cleaning, and the attempt to close the 'hole' in the North Sea open to vessel-borne discharges of such wastes seemed feasible during a transitionary period granted for the refurbishing of port and tanker operations of the prospective parties to the 1962 amended treaty.

Between drafting the 1962 amendments and subsequent moves to alter the oil pollution treaty, however, North Sea offshore oil exploration activities began in earnest, and the delivery of oil supplies by means of tankers in the area, apart from the few pipelines directly connecting selected production sites to user countries, intensified the familiar problems of transporting oil without at the same time causing further deliberate pollution of the sea by waste oils from tankers. If tankers were to be prohibited from discharging oily wastewater in excess of the effluent quality (100 ppm) described in the 1962 amended treaty, then waste reception facilities and other tanker modifications indicated costly investments and multi-company harmony on a scale not considered feasible by the industry.

In addition to the problems of scale in meeting the total prohibition ban of the 1962 amended treaty, the industry also faced the possibility of strong public pressure to include other types of marine pollution in the next round of international negotiations to control shipping pollution. Significantly, the North Sea lies at the heart of the world petrochemical trade and correspondingly receives the preponderant share of noxious liquid cargoes discharged by product and chemical tankers at sea (see Fig. 8).

At the 1973 Conference, the North Sea was inexplicably omitted as a special area under the world's first comprehensive marine pollution Convention. A sense of urgency then had compelled the designation of several special areas in the world wherein higher conservation standards would be regarded as a matter of priority, but the North Sea does not appear in this category. In fact, it is the only major maritime region which fares less favourably in the 1973/78 Convention. Other areas – the Baltic, Black Sea, Mediterranean, Red Sea, and the Arabian/Persian Gulf – have either retained or

Figure 8. World Distribution of Noxious Liquid Cargoes by Sea

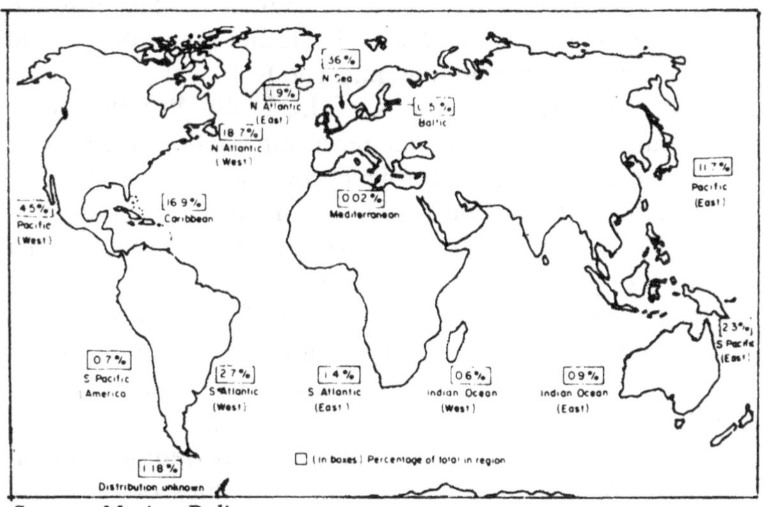

Source: Marine Policy

upgraded their claims to special protection due to geographical, ecological or shipping conditions. Regional arrangements have been taken to implement the 1973/78 Convention standards for these special areas, but unfortunately the North Sea, which needed the greatest protection due to subsequent developments, was shunted to the background.

By reverting to a minimal status, the North Sea has been laid open to deliberate pollution by ships, albeit not the most toxic wastes for which total retention rules apply. Based on the industry's historic influence on the policies of major European maritime powers (as we have seen in previous negotiations), it is also apparent that less stringent marine control policies were initiated for general application, but which may have had the ulterior motive of both hiding and meeting the industry's difficulties in coping with the particular problems of the North Sea development.

Due largely to the oil and shipping industry's lavish publicity campaign in the mid-1960s, the 1969 amendments to the oil pollution treaty were framed to legalize the Load on Top discharge system for all tankers in service. At that time, the ideal of total prohibition embodied in the 1962 amended treaty was exchanged for what the industry claimed was a more practical and cheaper control method. However, such a concession proved futile, as industry checks revealed a disappointing compliance with LOT, and pollution reached unprecedented levels in the mid-1960s to early 1970s.[290] Secondly, industry champions of LOT also promised that they would voluntarily provide reception facilities for tankers engaged in short voyages for which LOT is not applicable.[291] But evidently, it is a flimsy proposition to count on the promises of prospective offending parties when the International Chamber of Shipping has now argued against closure of the North Sea to operational pollution,[292] thus confirming that commercial development in the area has not been accompanied by a corresponding responsibility of the industry to protect the environment.

Waste reception facilities at ports are required of all parties to the 1973/78 Convention, and ships of countries accepting the treaty would have to operate an improved LOT discharge system or to retain all slops onboard for alternative disposal or use. But apart from the resistance on economic and technical grounds, it seems likely that port reception facilities would continue to be provided only according to the reduced requirements of ships which had washed out in areas yet open to routine discharges. Indeed, special area authorities in the Mediterranean seeem to count on such a transfer of pollution.[293]

Meanwhile, chemical tankers would replicate the LOT discharge system with some modifications on standard discharge rates, depth-of-water and distance-from-shore rules intended also to dilute the wastes in the receiving sea area. It is only when laden with chemical residues from Category A (highly toxic and bio-accumulative) substances that parties to the 1973/78 Convention must require their ships to retain all dirty washings and residues for disposal into port facilities. Outside of the Baltic and the Black Sea, the two special areas for Annex II regulations, discharges of residues of Categories B, C and D wastes (less toxic than Category A) are allowed according to an ascending degree of permissiveness.

The procedures for chemical tanker discharges selected by the International Maritime Organization are based according to the 300-second duration time rather than more stringent 10-second minimal manifestation of the noxious wastes in the ship's wake.[294] Taking the average contents of five tanks in a chemical tanker, it would take a sea journey of some 220 miles to control pumping when duration time is set for 10 seconds; whilst the same tanker may complete routine overboard washings through only 50 miles on the

Figure 9. North Sea offshore rigs and Western European tanker routes

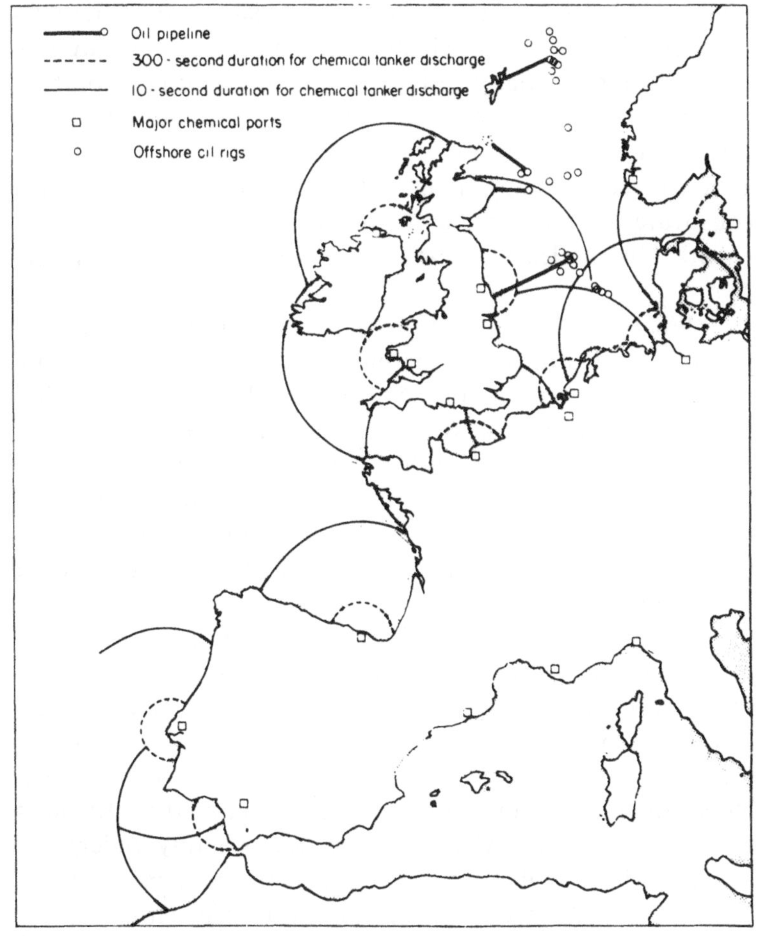

Source: Marine Policy

300-second concentration time. Thus, even with the proposed restraints foreseen in the 1973/78 Convention, chemical tankers, especially within the closely-situated North Sea ports, may continue to empty their tanks of chemical residues at sea without extending their journeys or relying on shore facilities for their remaining slops (see Fig. 9).

While the probable toxic effects of discharged chemical wastes would be greater for the 300-second duration time, it was nonetheless suggested that the hazard seems pronounced only for about 10 Category B substances with a moderate rating, e.g. phenol. [295] Overall, it was explained that the safety factor may be more considerable, and conditions for most other noxious chemical wastes, more favourable. [296] It is also widely believed that adult or juvenile fish, on which toxicity tests were mainly based in laboratory experiments, could be relied upon to keep out of harm's way or else metabolize or excrete the pollutant, but there is an unvoiced assumption here that lower organisms, which constitute a vital element in the food chain at sea, may be regarded as the least of our hostages to the fortunes of pollution.

What underscores the problem is the controversy on the biological assessment of marine pollution and the oceanographic character of the North Sea. Progressively richer harvests have been assessed by North Sea fishery officials. [297] Such statistics contrast with the action by Scandinavians to ban selected species farming due to fears of overfishing, and highlight the reasonable doubt that progressively richer harvests of North Sea fish may be attributed to modern fishing techniques and more numerous trawlers than to the inconsequential effects of pollution. A more subtle indication of chronic harm may have been provided by a 20-year monitoring study of North Sea plankton, the basic staple of life at sea, which shows definite signs of decline and progressively delayed seasonal reproduction. [298] Such changes are attributed open-mindedly either to natural climatic influences or to pollution. But what biological assays hardly underscore is the obvious – if North Sea marine life may be inhibited by natural influences, chronic pollution can hardly be expected to improve its condition and habitat.

Another argument against closing the North Sea to deliberate pollution has been its geographic and oceanographic characteristics. The currents passing through the Skagerrak and Kattegatt seem to promote a cleansing effect on the waters of the Baltic at great depths, which tend to suggest that the North Sea may have a higher tolerance of pollution than can be endured by semi-enclosed seas. But further studies by experts tended to show that the North Sea, rather than ventilating the waters of the Baltic, seems to have become an additional source of pollution in the latter.[299]

It remains to be seen whether official and public concern may lead to a reconsideration of preventive measures for the deliberate pollution by oil and some chemical wastes from vessels traversing the North Sea, since various international and voluntary schemes hardly meet the problem. Reclaiming the North Sea as a special area, and an amendment to the 1973/78 Convention to this effect, would result not only in a viable movement to upgrade port and tanker operations to the similar degree which has begun in other important waterways, but it would also mean a long-term and practical conservation of North Sea resources and environment and a sheer reckoning with a privileged past.

CONCLUSION

In our survey of the history of oil pollution control, it is possible to identify several factors which have influenced policy, either to promote or to constrain the adoption of anti-pollution measures. Such factors can serve equally well as a framework for analysing any environmental problem requiring social action. Like any environmental problem, oil pollution must be regarded as synergistic, for even as the factors influencing its policy aspects operate in confluence, so does its desired solution require interdependent and not isolated action. Overall, the valuable lessons learned after six decades of experience with oil pollution need to be integrated when approaching other problems involving waste management and conservation.

Influences on Policy

Any environmental problem has several factors which determine the fate and formulation of social intervention policy. Fig. 10 illustrates the factors which either promote or limit such response. Whilst it is not the purpose of this book to propose a theoretical framework, nonetheless the history of oil pollution control illustrates the confluence of these factors in the outcome of attempts to solve the problem.

The evidence tends to show that the response of governments and the oil and shipping industry will be conditioned by their perception of and the importance they attach to the factors of consequence to them.

Figure 10. Factors Influencing Environmental Policy

Factors Promoting Policy	Factors Constraining Policy
1. Developments in the use and transport of oil	1. Cost of anti-pollution control outweigh perceived benefits
2. Increased pollution and public pressure; evidence of environmental damage	2. Minimal pollution and public apathy; evidence of acceptable risk
3. Inadequate anti-pollution measures and the initiative of government officials and industry leaders	3. Adoption of new measures to prevent, control, mitigate, or compensate for pollution
4. Unilateral government measures ahead of or in excess of international standards	4. Political reluctance to upset the status quo

Positive Factors

Developments in the use and transport of oil tend to promote a policy of controlling pollution arising from these activities. Pollution due only to operational (deliberate) pollution by ships accounts for about 30% of oil pollution, or a loss of around 1 million tons of oil a year. The adage, 'Where there's money, there's muck,' becomes apropos when considering that an increase in the number of oil shipments and oil ships has a direct proportion to the incidents of pollution in an area.

An increase of pollution due to chronic or accidental discharges provokes an articulation of public grievances over damage to coastal resources, amenities and wildlife. Supported by scientific evidence on the extent or else the more subtle long-term consequences of pollution, public agitation will flow through the media and appeals will be made to governments officials and the industry.

Whatever arrangements were in place prior to the increase of pollution and public agitation, would appear inadequate and open the door to the search for new administrative, legal or voluntary anti-pollution measures. Initiatives by government officials and industry leaders generally facilitate the creative endeavour towards a solution.

Finally, the threat or actual promotion of a unilateral move by a state outside of international arrangements also has the tendency of compelling other states to reconsider existing international regulations.

Negative Factors

On the other hand, there are negative factors which delimit or constrain environmental policy. These also consist of economic, social, technical and political considerations.

Obviously, the industry would be reluctant to invest in anti-pollution measures where they have not previously costed these into their capital or operational expenditures. Where the costs would fall on the companies of parties to a Convention, leaving other companies in non-Convention states free of similar obligations, the issue of unfair competitive advantage complicates the picture. Furthermore, some of the benefits of pollution control appear as intangibles – the aesthetic, the recreational value, and the moral principle – against the tangible benefits of commercial and technological growth. However this argument is more apparent than real, for any immediate gains in economic progress at the expense of environmental and moral principles prove illusory in the long run. It is a tragic commentary on our contemporary society's values that immediate economic advantage is often preferred regardless of other consequences.

Minimal pollution, public apathy, or scientific conclusions on the acceptable risk of pollution in an area obviously would preclude any initiatives and render unacceptable inquiries from other quarters to join in anti-pollution action.

Ironically enough, the adoption of any measure to prevent, control, mitigate, or compensate for pollution has the initial effect of parrying off criticism or public clamour on pollution. No matter how inadequate these measures may turn out, they have the effect of palliatives and the appearance of action, to 'show people we have done something'. For example, in that we have established the procedure and machinery for cleaning up oil spills, compensating victims, automatically monitoring discharges from tankers, we may have tacitly admitted failure in preventing pollution at the source.

Finally, whatever the nature of pollution may be in a certain country, a conservative attitude of government officials with regard to national and international interests would hamper efforts to upgrade environmental protection by means of better national or international regulations. Although attitudes have improved, states usually place a lower priority on environmental problems than such national interests as trade and security.

Policy Assessment: An Overview

In the early 1920s, oil pollution of ports and coastal areas become so widespread and obnoxious that various governments passed local and national measures and considered international regulations. National laws by Britain, the United States, and other major maritime countries, covered only their own national fleets and territorial waters against operational pollution arising from operational pollution by ships. But these did not provide for accidental pollution or the compensation of victims of pollution, and they were not entirely satisfactory.

It became apparent that isolated national efforts had to be complemented by internationally-agreed control standards. To find more effective means of controlling oil pollution and to attract international action, the United States government initiated various studies and diplomatic negotiations culminating in the 1926 Washington Conference of experts, which drafted an oil pollution agreement. The 1926 draft Washington convention was widely misunderstood and delayed by ill-fated turn of events, so it was ultimately abandoned.

But the 1926 draft convention created certain precedents and pointed to promising means of control. Firstly, governments would have established wider zones off their coasts wherein ships of parties to the treaty were to be prohibited from discharging 'persistent oils' in excess of permissible limits and were to be subject to flag state jurisdiction. Even more significant than the zone system, which almost everyone accepted as a mere public show-piece or stopgap measure, the 1926 draft treaty contained various incentives which would have attracted greater conservation techniques. Future parties to the treaty would have had to remove every disability preventing tankers and ships from equipping for or practising retention of oily wastes onboard. These incentives included the waiver of customs and canal toll fees for oily slops, exemption from tonnage admeasurement due only to oily water separating equipment or tanks containing oil slops, and other penalties on ships wanting to retain oily wastes onboard. It was hoped that in such a manner shipowners and tanker operators would be encouraged to modify their operations or their ships. But technical, economic and political difficulties militated against the implementation of the proposed measures. The Great Depression wiped away any hopes of financing the voluntary installation of separators on ships or of more widely providing port reception facilities for ships' oily wastes.

Meanwhile, American enthusiasm for international measures diminished after the good results brought about by its 1924 Oil Pollution Act and the voluntary measures initiated by the international maritime industry. Furthermore, an American scientific experiment in 1927 dealt a blow to the belief held by those who had previously advocated strong controls that crude and fuel oil could persist indefinitely in the marine environment. Other important maritime powers who attended the 1926 Conference delayed and obstructed the ratification of the 1926 draft treaty. Influential officials, especially some of the British and American negotiators at the 1926 Conference, departed from the scene. In some countries, oil pollution had decreased due to shipbuilding innovations and the zone system voluntarily practised by the industry, as well as the passage of national legislation.

Mainly to placate domestic critics and to guard against special measures that could upset law of the sea negotiations, the British government revived international interest through the League of Nations in the mid-1930s. The 1935 draft League convention, framed by an ad hoc Committee of Experts, was stylistically more elegant than the 1926 treaty, but in terms of the proposed controls it was actually a weaker instrument. Parties to the League draft would have been asked to introduce a zone system only. The earlier proposals on incentives for retention onboard and equipment of ships, and new moves to call for port reception facilities, were shunted to the status of recommendations in the draft Final Act rather than included in the obligations of the draft treaty.

Meanwhile, governments, pre-occupied by other problems of the pre-war era, exhibited uneven interest in controlling oil pollution by inter-governmental action. If they were not keenly enthusiastic, at least neither were they adamantly obstructive as in the 1920s. Strangely enough, had the British government persisted in their initiatives it might have been possible to start a limited system of control. According to the League survey of states' opinions, some governments were even in favour of equipping new ships for onboard retention of oily wastes. Significantly also, shipowners had begun to install separators or otherwise practised the voluntary zone system in areas wider than covered by national legislations. But British officials professed to see various obstacles to the League convention.

After the outbreak of the Second World War, the hostilities put paid to the efforts through the League. Official documents on the subject later perpetrated the convenient excuse that Germany, Italy and Japan (the Axis powers) had 'obstructed' early efforts to control oil pollution. In fact, much of the responsibility could be attributed to the

American and British governments which did not pursue their initiatives.

The Second World War introduced special measures to conserve and retain oily wastes due to strategic and safety considerations. But these measures were discontinued after the war, and shipowners reverted to normal operations, including operational pollution by ships at sea.

At the request of the British government, action within the United Nations Transport Commission was abandoned in favour of Britain sponsoring an international conference on oil pollution. Out of this conference came the 1954 Oil Pollution Convention, which became the world's first working treaty to control oil pollution from the normal operations of ships at sea. Apart from enshrining the worldwide system of oil-free zone beyond territorial waters, the 1954 Convention indicated new ways of preventing shipping pollution. Parties to the treaty were obliged to provide port reception facilities for non-tankers.

It was understood at the time that another conference would be called within three years to assess the working of the Convention. But the second conference was delayed due to the late ratifications of the treaty and due to the work of IMCO, which would sponsor the next conference, with the safety of life at sea convention. Other problems of a political nature arose, with some important states like the United States delaying ratification of the 1954 Convention only in anticipation of the opportunity to revise its provisions at the next conference.

The 1962 amended Convention improved upon the original treaty by requiring total retention of oily wastes in all large ships built after 1967, when the revised agreement entered into force. The 'total prohibition' clause for new large ships was regarded by environmentalists as a major breakthrough and as 'the signpost of the future'. However, before it could be implemented, the provision was sidetracked by the introduction of the Load on Top system, which the major oil companies foisted on the public. Largely based on the pledges of its commercial sponsors, LOT was welcomed as an immediate practical substitute in lieu of the long-term ideal of port and tanker modifications implied in the 1962 total prohibition clause. The LOT system offered the semblance of reform without the substance, and its disappointing performance, despite the 1969 amendments, lent impetus to the movement for more effective control measures.

From the mid-sixties to the early seventies, unprecedented levels of pollution due to operational discharges and accidents involving large tankers created a climate of opinion favourable to a comprehensive policy on oil pollution. The legacy of the *Torrey Canyon* and other pollution disasters inspired new efforts in preventing, controlling and mitigating pollution, as well as compensating victims for pollution damage. These efforts involved international and regional agreements, national programmes, industry codes, and innovations in clean-up systems, port and tanker modifications, and insurance funds.

Meanwhile, greater awareness that industrial progress and the quality of life could not be sustained without better policies on major pollution problems presented opportunities on a global scale to raise public consciousness. New global, regional and state agencies were created to focus on solutions to important environmental issues; scientific and popular literature provided much-needed information; and pressure groups urged more responsible attitudes and conduct towards environmental problems.

The 1973/78 Marine Pollution Convention, as supplemented by other maritime standards, represented a last ditch effort to eliminate deliberate vessel-source discharges of oil at sea by the 1980s, and an innovative attempt to control accidental pollution by oil and to minimise discharges of other noxious liquid cargoes, sewage and garbage. However, as we have seen, these arrangements for marine pollution control yet fall short of their stated objectives due either to inherent defects in the provisions or the non-implementation by governments and the industry.

Management and Personal Ethic

An inevitable lag between the declared aims of governments, the industry and individuals, and their performance in solving the oil pollution problem, basically stems from lack of co-operation and the inherent human characteristic to maximise gains and to minimise obligations. Although the environmental problem is holistic and pervasive, the social response thus far has been to compromise, to accept palliatives, and to shift the responsibility onto others.

In practical terms, such an approach translates into real or potential conflicts among the various interests concerned and slow progress in resolving waste and environmental issues. Governments with more serious pollution problems and more articulate environmental lobbies show a tendency to welcome internationally-agreed compromises or to pass unilateral measures exceeding international standards; whilst less-developed states fail to meet their treaty obligations or opt out of international agreements due to inadequate facilities and resources. The oil and shipping industry has yet to convincingly demonstrate that it is not motivated by the unbridled quest for economic advantage. Individual desire for change invites a personal quest for an alternative lifestyle – simpler, more frugal, not hedonistic – and the admirable leadership of some people to change their values after the 1970s oil crisis has yet to be matched by the complacent majority.

The history of oil pollution control shows that the solutions to the problem have always been available. Leading experts in the problem acknowledge that it is well within the capability of governments and the interests involved to implement the various technical and legal provisions for the problem. In many ways, tremendous improvements have been made, and there is hope in the optimistic assessment by official observers like Thomas A. Mensah, Assistant Secretary-General of IMO, that satisfactory results have been obtained from the implementation of national and international arrangements.[300]

Oil pollution provides a model for environmental policy-making in the sense that the prevailing opinion maintains the satisfactory settlement of the problem, and secondly, the lessons gained from a study of its resolution are applicable to other environmental questions. Even as the discovery and use of oil propelled progress in this century, its concomitant waste management and conservation have provided the opportunity to gauge the present transformation of our modern society's values and lifestyles. As Noel Grove put it, 'If the first gift of oil was the gift of progress, its next may be the gift of concern – for the way we live, for our diminishing resources, and for our earth.' [301]

Conditions may alter opinion, and the replication of historical principles do not always guarantee their integration by policy-makers into present issues. The variables in the environmental phenomenom and in social activity continue to interest contemporary analysts, who offer partial guidance on what steps may be taken towards reconciling the problem. Briefly, their suggestions generally express reliance upon the possibility of transforming values and institutions through exhortations to apply reason, altruism and aesthetics.

Ironically enough, a renewal of the human spirit and triumph over circumstances can only be achieved by admitting the limitations of human creativity, resources and strength. The third horseman of the Apocalypse is eco-catastrophe, and it seems appropriate in this troubled time to admit that to a large extent we have failed in our God-given stewardship over all life upon the earth and the skies and the seas. The only reliable choice in renewing the human spirit and obtaining victory over whatever circumstances is the recognition of a divine order in nature and in society. This is not an invitation for complacency but a justification of the liberating faith in God's promises. Though we may possess the desire and the capability to solve environmental problems, or any problem for that matter, the certainty is that, as Christ said, apart from Him we can do nothing.

NOTES

1. For the wartime instructions against oil pollution, see Public Record Office document M[inistry of] T[ransport] 9/15/71 (M[arine Department] 8981/18) and *The Journal of Commerce and Shipping Telegraph* (later referred to simply as *Journal of Commerce)*, 26 August 1919.

2. See 34 and 35 Henry VIII.

3. For the oil problem in the United States, see U.S. Senate 68th Congress 1st Session, *Pollution of Navigable Waters: Hearings before a Subcommittee of the Committee of Commerce on S. 42, S. 939 and S. 1388, January 9, 1924* (GPO: Washington, D.C., 1924), p. 11. This document will be referred to later as the '1924 Senate Hearings.' See also letter of G.W. Booth, U.S. National Board of Fire Underwriters, to the British Home Secretary, MT9/1521 (M 4520/21) and *Journal of Commerce*, 22 January 1921.

4. British officials in the 1920s professed to have found 'precedents' for the proposed Act in regulations passed by Bermuda (6 January 1921), South Australia (20 April 1921) and South Africa (1916), see MT9/1571 (M 12815/21 and M 13235/21). A British lawyer, Dr. W.R. Bisschop, was encouraged by Charles Hipwood, head of the Board of Trade's Marine Department, to bring the problem before the 1924 International Law Commission conference in Stockholm and was provided with information noting the three 'precedents' to the 1922 British Act on oil pollution. Accordingly, Bisschop reported these 'precedents' in his articles for *The Manchester Guardian* (31 July and 23 October 1924). C. John Colombos, in his classic textbook on maritime law, continued the error of regarding the South African and South Australian regulations as national laws, and he also failed to mention the third case of Bermuda, see *The International Law of the Sea* (London: Longmans, 1967), p. 431. These 'precedents' however are harbour regulations rather than national law, and there were numerous instances of similar port oil pollution regulations.

5. The documents on the 1921 British negotiations leading to the world's first oil pollution legislation are found in MT9/5348 (M 1060/50) and MT9/1571 (M 22279/21). I am also grateful for the background information provided by Mr. J.R. Willis, who was an assistant to Sir Charles Hipwood at the time.

6. See 12 and 13 George VI, Chapter 39.

7. The secret Board memo is found in MT9/1654 (M 14386/22).

8. Undated Board memo, MT9/1629 (M 15619/23).

9. 174 Commons, col. 189 (27 May 1924).

10. See Joint Public Res. No. 65 – 67th Congress 2nd Session (2 July 1922).

11. Public Law No. 238, 68th Congress 2nd Session (7 June 1924).

12. See *Oil Pollution in Navigable Waters: Report to the Secretary of State by the Interdepartmental Committee on Oil Pollution of Navigable Waters, March 13, 1926* (GPO: Washington, D.C., 1926), later referred to as 'Interdepartmental Committee Report'.

13. In 1927, the master of the s.s. *Inverarder*, a British vessel, was caught, jailed and fined in San Pedro, California, for discharging oil within the restricted area; see PRO F[oreign] O[ffice] 371/12045 (A[merican Department] 6600 and A7215/224/45).

14. See Board correspondence with U.S. officials in MT9/1567 (M 11727/22), MT9/1625 (M 9031/23), (M 13594/23) and MT9/1726 (M 7935/26).

15. See U.S.-British correspondence and other memos in MT9/1567 (M 11727/22) and MT9/1625 (M 9031/23).

16. See Hipwood's letter to port officials, shipping and oil companies, July 1923, MT9/1625 (M 9031/23) and minutes of the round-table conference with Frelinghuysen, held at Board headquarters, on 20 July 1923, ibid.

17. Report by J.R. Willis, 'Inventions for Separating Oil from Water,' 27 October 1921, MT9/1571 (M 17397/21) and also the 1925 White Paper, *Oil in Navigable Waters: Reports as to the Extent of Oil Pollution Round the Coasts and Reports on Oil Separators* (later cited as 1925 White Paper).

18. FO 371/9609 (A 3452/211/45).

19. FO 371/9609 (A 4591/211/45).

20. FO 371/10651.

21. See Interdepartmental Committee Report in note 12.

22. See *Pollution by Oil of Coastal Waters of the United States*. Preliminary Report (GPO: Washington, D.C., 1923), pp. 5-6.

23. Ibid. The British noted that their investigations came to similar conclusions, see memo of Willis, 30 October 1923, MT9/1625 (M 16653/23).

24. Preliminary Report, op. cit. note 22, pp. 47-49.

25. 1924 Senate Hearings, op. cit., note 2, p. 79.

26. See statement of Robert F. Hand, American Petroleum Institute, in 1924 Senate Hearings, ibid., pp. 79-81.

27. For a description and illustration of the *Charles Pratt* experiment by a former company executive, see James E. Moss, *Character and Control of Sea Pollution by Oil* (Washington, D.C.,: American Petroleum Institute, 1963) pp. 56, 74-75 and his article in Donald W. Hood (ed.), *Impingement of Man on the Oceans* (New York: John Wiley & Sons, 1971), pp. 393-394 and 396.

28. See Appendix 5 of the Interdepartmental Committee Report, op. cit., note 12, Dr. Stroop later became a consultant of the American Petroleum Institute.

29. Interdepartmental Committee Report, op. cit., note 12, pp. 4-5 and FO 371/11188 (A 2254/316/45).

30. Committee Report, ibid., p. 102.

31. See Stanley Hornbeck's remarks to C.J.W. Torr, First Secretary of the

British Embassy in Washington, D.C., FO 371/14295 (A 4378/3197/45).

32. Confidential records of the British government for this period show that the environmental groups wrote numerous of letters, sent annual or frequent petitions and complaints or clippings of news reports on oil pollution. Some of these complaints seemed ironic, as an irate reader of *Angler's News* wrote, 'Ten years of this evil and there will be neither seafowl nor wildfowl to see, let alone to shoot!' See RSPB *Bird Notes and News* XI:2 (Summer 1924).

33. The May 1925 petition to the Prime Minister was cited in the 1925 White Paper, op. cit. note 12, p. 3. Britain's worst oil pollution then fell between the Isle of Wight and the north Kent coast, which included the English Channel and the ports of London and Southampton

34. See FO 371/19613 (W[estern Department] 1361/76/50).

35. Lord Montagu to Hipwood, 1 July 1925 (MT9/1657 (M 11503/25), *Lloyd's List and Shipping Gazette* (later referred to as *Lloyd's List),* 14 May 1924. A resolution by the RSPB Secretary, Mrs. Frank E. Lemon, was warmly endorsed however by delegates to the International Nature Protection Congress, Paris, 1924, *Bird Notes and News* (Summer 1924), p. 10. See also *Lloyd's List,* 23 June 1924, and *Manchester Guardian,* 23 October 1924.

36. FO 371/19615 (W 9589/75/50).

37. Minutes of A. Hopper, 27 July 1925, and Admiral Munro, 23 March 1925, in MT9/1694 (M 11503/25).

38. U.S. State Department Press Release, 8 April 1926, in Fo 371/11188.

39. Minutes of a round table conference at the Board, involving oil and shipping companies and port authorities, 3 May 1926, MT9/1729 (M 7935/26).

40. Ibid., p. 2. In fact, American officials never contemplated going go far as to close their ports to foreign ships; Dr. Arthur N. Young's letter to the author, 12 March 1973.

41. Instructions to the British delegation, see FO 371/11188 (A 2786/316/45). See also Board memorandum in the same file, document (A 2703/316/45).

42. Unless otherwise indicated, the remarks and data are taken from the verbatim proceedings of the conference, *Preliminary Conference on Oil Pollution of Navigable Waters, Washington, D.C., June 8 to 16, 1926* (GPO: Washington, D.C., 1926).

43. Dr. Young's letter to the author, 12 March 1973.

44. See Moss, op. cit., note 27, pp. 73-82. See also testimony of M.P. Holdsworth before a Select Committee of the House of Commons, *Coastal Pollution Report, Minutes of Evidence, Appendices and Index* (London: HMSO, 1968) p. 128.

45. See Report of British Delegates to the Board of Trade, no date, copy in FO 371/11189 (A 3623/316/45). See also Note by the Board of Trade, 1933, in MT9/2285 (M 10726/33).

46. Oil company experts are reluctant to acknowledge that prior to the 1960s the procedures for LOT were known and implemented on a smaller scale because of their claims about its novelty in 1964, see Part Five.

47. Britain, Canada, the U.S., Denmark, France, Italy and Norway voted for the zone system as only an interim cure. Germany, the Netherlands and Sweden voted for it as a permanent solution. Belgium was absent, and Japan wanted it referred back to the committee on zones.

48. Dr. Young's letter to the author, op. cit., note 43. According to Dr. Young, only one word was changed at the plenary session.

49. Colombos and Mouton, writers of international law textbooks, mistook the terms embodied in the Final Act of the 1926 Washington conference with that found in the 1926 draft convention, and they give a wrong interpretation of the conclusions of the conference, see Colombos, op. cit., note 4, p. 431 and W.W. Mouton, *The Continental Shelf* (The Hague: Martinus Nijhoff, 1952), p. 162.

50. Proceedings of the 1926 conference, op. cit., note 42, p. 164.

51. RSPB letter of 28 August 1934, FO 371/18498 (W 7804/1324/50).

52. 1926 conference proceedings, op. cit., note 42, p. 164.

53. 1926 draft convention, Article VI, and Final Act of the Washington conference, Recommendations 7, 8 and 9, ibid.

54. Moss, op. cit., note 27, p. 56, and his 1971 work, pp. 393-394.

55. 1926 conference proceedings, op. cit., note 42, p. 358. Hipwood also withdrew a resolution calling for the provision of port facilities, ibid., p. 394.

56. Indeed Dr. Young and Hipwood never contemplated the possibility of raising the question of total prohibition, see their exchanges of 14 January 1927, FO 371/12044 (A 716/224/45).

57. *Gaceta de Madrid,* 9 September 1925, No. 252, p. 1471.

58. Minute of 7 October 1925, MT9/1694 (M 15994/25).

59. Ibid., Willis's minute of 6 October 1925.

60. Ibid., copy of British Note to the Spanish Foreign Minister, 23 Dec. 1925.

61. Decree No. 14,354, copy in FO 371/12712 (W 9550/9550/36).

62. Ibid., FO Memo of Harry Jones, 27 Oct. 1927. Portugal wanted the widest extent of territorial waters at the 1930 Hague conference.

63. FO 371/12712 (W 11730/9550/36), Gerald H. Villiers to Sir L. Carnegie in Lisbon, 4 January 1928.

64. FO 371/12712 (W 10278/9550/36) Charles Grimshaw to FO, 3 Nov. 1927.

65. FO 371/11188 (A 316/316/45).

66. Ibid., minute of 12 February 1926.

67. See Trinidad and Tobago Oil in Waters Colony Ordinance, 1951.

68. FO 371/11188 (A 316/316/45) Grindle's letter of 14 January 1926.

69. FO 371/24270 (A 301/301/47).

70. See ADM[iralty] 116/3926 (M 395/36).

71. ADM 116/3926 (M 287/36).

72. See FO 371/24270 (A 2645/308/47) and (A 2975/308/47); also E.W. Ravenshear's FO letter to W. Monson of the Colonial Office, 24 April 1940, in (A 2820/308/47).

73. United Nations Legislative Series, *Laws and Regulations on the Regime of the High Seas,* ST/LEG/SER.B/1 (11 January 1951) p. 47.

74. See Article 21 of the Submarine (Oil Mining) Regulations, 22 May 1945, of Trinidad and Tobago (Government Notice No. 87) in UN series, ibid. Mouton (op. cit. note 49, p. 169) is wrong to state that the Orders-in-Council gave rise to the drilling regulations because the latter was prepared first, even though they were not implemented until 1945.

75. FO 371/22850 (A 2266/301/47) letter to CO Undersecretary Boyse, 24 March 1939.

76. Cmnd. 6400 (London: HMSO 1942), Treaty between his Majesty in respect of the United Kingdom and the President of the United States of Venezuela, relating to the Submarine Areas of the Gulf of Paria, Caracas, February 26, 1942.

77. Cited by Mouton, op. cit. note 49, pp. 169-170.

78. See Convention on Civil Liability for Oil Pollution Damage Resulting from Exploration for and Exploitation of Seabed (Mineral) Resources, 1976, in Misc. No. 8 (1977), Cmnd. 6791, and *International Legal Materials,* Vol. 16, 1977, p. 1451. The 1976 Convention holds the drilling operators liable for oil pollution damage; liability is strict and can be limited to 30 million Special Drawing Rights (SDRs). Victims of oil pollution damage from offshore drilling in the North Sea may also avail of a non-governmental and voluntary scheme offered by the oil companies, called the Offshore Pollution Liability Agreement (OPOL). OPOL covers operators' liability in Britain, Denmark, France, Ireland, the Netherlands and Norway, see text in R. Churchill et al (eds.) *New Directions in the Law of the Sea* Vol. IV, (New York: Oceana Publications, 1977).

79. For studies on the British Commonwealth, see J.E.S. Fawcett, *The British Commonwealth in International Law* (London: Stevens and Sons, 1963); J.C. Beaglehole, 'The British Commonwealth,' in *New Cambridge Modern History* (Cambridge University Press, 1968); and Nicholas Mansergh, *Survey of British Commonwealth Affairs: Problems of External Policy 1931-1939* (London: Oxford University Press, 1952). For a full description of this topic, see author's Ph.D. thesis, *The International Politics of Oil Pollution Control, 1920-1962* (London School of Economics and Political Science, 1975), pp. 87-97.

80. FO 371/11189 (A 4341/316/45).

81. Ibid.

82. FO 371/11189 (W 5918/316/45), minute of W. Eric Beckett, 10 November 1926.

83. For a table and description of the British amendments to the 1926 draft treaty, see author's Ph.D. thesis, op. cit. note 79, pp. 109-113.

84. FO 371/13544 (A 561/561/45) minute of C.E. Steel, 23 January 1929.

85. See author's Ph.D. thesis, op. cit. note 79, pp. 98-108; also U.S. Memorandum of 20 September 1926, FO 371/11189 (A 5045/316/45).

86. Letter to the author, 12 March 1973.

87. MT9/1841 (M 13456/26), minute of 5 November 1926.

88. ADM 116/2524 (M 1384/27), minute of 24 February 1927.

89. FO 371/11189 (A 6823/316/45).

90. For minutes of this meeting, see FO 371/12045 (A 2903/224/45).

91. FO 371/12045 (A 3119/224/45).

92. Atherton to FO Undersecretary, 16 November 1927, FO 371/12045 (A 4564/224/45) and FO 371/12045 (A 7018/224/45) for the FO despatches.

93. FO 371/12802 (A 1004/19/45). In reply to this, Dr. Young recalled that Campbell might have contacted the geographical division personnel before coming to see him, letter to the author, 12 March 1973.

94. FO 371/12802 (A 821/19/45).

95. FO 371/12802 (A 815/19/45).

96. FO 371/12802 (A 5099/19/45).

97. FO 371/12802 (A 5965/19/45).

98. The German shipowners views are found in FO 371/18498 (W 8175/ 1324/50.

99. Minute by J.V. Perowne, 16 November 1928, FO 371/12802 (A 7723/19/45). For the British-German exchanges, see FO 371/12802 (A 2155/ 19/45).

100. Minute of 26 October 1928, FO 371/12802 (A 7364/19/45).

101. FO 371/12802 (A 6688/19/45).

102. D.V. Stroop, 'Report on Oil Pollution Experiments: Behavior of Fuel Oil on the Surface of the Sea,' December 1927, National Bureau of Standards, copy in FO 371/12802 (A 1077/19/45).

103. Admiralty Hydrographer J.S. Edgell questioned the validity of these results on the basis of their own tests in 1934, because, as he noted, Dr. Stroop had used pure dry oil which was not the type normally discharged from ships at sea. Furthermore, he noted that the experiments determined the results only by inspection without instrumentation. Admiralty experiments, however, showed that oil can form globules even without showing up as a visible sheen. See FO 371/19613 (W 4103/76/50). A senior scientific officer with the Ministry of Agriculture and Fisheries, C.H. Roberts, also conducted independent experiments and came to the conclusion that fuel oil can drift almost indefinitely in the marine environment. See Robert's memorandum of 23 April 1934, ibid. This became the basis of the supplementary report of the British Government to the League of Nations questionnaire, FO 371/19614 (W 6728/76/50).

104. FO 371/12802 (A 3262/19/45) and (A 5781/19/45). Sir Ronald Campbell, in recounting the same events, said that his interest was due to that held very keenly by Robert Vansittart of the Foreign Office and by Campbell's own experience with oil pollution on seabirds in Dorsetshire; letter to the author, 24 July 1974.

105. Terence Allen Shone's interview with Paul T. Culbertson of the State Department, 31 May 1929, FO 371/13544 (A 3731/561/45).

106. FO 371/14295 (A 4378/3197/45).

107. For minutes of the Board meeting, see MT9/2008 (M 15893/29).

108. For the summary of replies from H.M. Coast Guard, coastal town councils, harbour officials and other bodies, see Charles Grimshaw's letter to Robert Vansittart, 16 October 1930, FO 371/14295 (A 6742/3197/45).

109. MT9/1857 (M 14597/28) and also 207 House of Commons Debates, 5s, col. 187.

110. *The Destruction of Birds by Oil Pollution,* copy in ADM 116/2524 (M 4837/28).

111. Cited in MT9/2285 (M 8072/33).

112. MT9/2008 (M 10091/29), see also previous section.

113. Miss L. Gardiner, RSPB Secretary, to the Board, 28 June 1929, MT9/2008 (M 6490/29); FO 371/14295 (A 4070/3197/45); FO 371/15138 (A 3210/2678/45); and FO 371/17329 (W 8939/7252/50).

114. *Lloyd's List,* 29 July 1931 and 255 HC Deb., 5s, cols. 2300-2301

115. Chamber memorandum of April 1937, sent to the Board and copy furnished the FO in FO 371/21248 (W 7850/469/98).

116. MT9/2285 (M 10726/33).

117. FO 371/7329 (W 9937/7252/50).

118. Minute of 10 October 1933, FO 371/17329 (W 12231/7252/50).

119. Minute of 1 January 1933, FO 371/17329 (W 9937/7252/50).

120. FO 371/17329 (W 8849/7252/50). See the reply of Mrs. Rosalia Edge, Chairman of the New York Conservation Committee, to McCormick-Goodhart, 10 November 1934, FO 371/18498 (W 10250/1324/50). The President of the Massachusetts SPCA, Dr. Francis H. Rowley, published a pamphlet, *An International Appeal* (Boston, 1935); a resolution was passed at the 1934 Ornithological Congress in Oxford, FO 371/19514 (w 7153/76/50); and the Portuguese Marine Industry also received representations from their local SPCA, copy in FO 371/20477 (W 2279/110/98).

121. See Acts of the Conference for the Codification of International Law at the Hague from March 13 to April 12, 1930, vol. III, minutes of 2nd Committee on Territorial Waters, League of Nations C 351 (b); M 145 (b) 1930 V, Geneva, 19 August 1930. Also cited by the Admiralty in ADM 116/2524 (M 2579/30' Compare this assessment, based on primary government records with the confused interpretation given by Mouton, op. cit. note 49, p. 163.

122. FO 371/14295 (A 4070/3197/45).

123. FO 371/17329 (W 11219/7252/50) and (W 12231/7252/50).

124. See FO 371/17329 (W 9103/7252/50), (W 12668/7252/50), and (W 1736/7252/50).

125. FO 371/18498 (W 1324/1324/50).

126. FO 371/18398 (W 4166/1324/50).

127. For the secret Admiralty process for treating oily wastes onboard ships, see MT9/5071 (M 4435/36). The process was developed in 1929 and became more widely used after the Second World War.

128. Quotes from Lord Cecil's remarks at the Lords debates on 22 February 1934 were added to the memo for emphasis, see 90 HL Deb., col. 946, and Board nemo, op. cit., note 126.

129. Shuckburgh's minute of 30 August 1934, FO 371/18498 (W 7804/1324/50).

130. FO memorandum, 15 August 1934, FO 371/18498 (W 7281/1324/50).

131. FO 371/18498 (W 5774/1324/50).

132. FO 371/18498 (W 5213/1324/50).

133. FO 371/18498 (W 7822/1324/50).

134. Ibid., Shuckburgh's minute of 30 August 1934.

135. FO 371/18498 (W 8421/1324/50) and interview with Sir Geoffrey Shakespeare, 1975. Sir Geoffrey recounted that he sponsored an essay contest in schools in his constituency on the problem of oil pollution.

136. Shuckburgh's minute of 24 September, 1934, FO 371/18498, ibid.

137. League of Nations A.II/P.V.2.1934, 15th Ordinary Session of the Assembly, 2nd Committee, 2nd Meeting, 12 September 1934.

138. See FO (371/19613 (W 76/76/50) for the committee's confidential proceedings.

139. Ibid.

140. Ibid., FO note of December 1934. During the compilation of this survey and other work of the League Committe, Grimshaw did not wish the proceedings made public, for the discussions would have exposed British lack of commitment to anything more than the zone system, see FO 371/19613 (W 4103/76/50).

141. For the questionnaire, see League of Nations C.527.1934.VIII (Geneva, 8 December 1934); for the replies, see League of Nations, A.20.1935.

VIII (1 August 1935) and Addendum (1 August 1935).

142. FO 371/19613 (W 9399/76/50), Cleminson to Grimshaw, 18 October 1935.

143. See FO 371/19613 (W 806/76/50) and (W 4103/76/b0) for a copy of Roberts' tests and the discussions between the latter and Grimshaw. See also Roberts' article in the *Journal of the International Council for the Exploration of the Sea* (1926). For the Admiralty tests, see FO 371/19613 (W 4103/76/50), copy of Edgell to Grimshaw, 8 November 1934, see also op. cit. note 103.

144. FO 371/19613 (W 811/76/50, Grimshaw to FO, 13 May 1935.

145. Ibid.

146. Ibid. See also Admiralty's views in FO 371/18498 (W 8175/1324/50).

147. League of Nations A.II/1./1935 (Geneva, 11 September 1935), 16th Ordinary Session of the Assembly, Second Committee, Work of the Organization for Communications and Transit. Grimshaw called this passage 'an exaggeration,' see MT9/2501 (M 5008/35).

148. See S.H. Phillips to Board Under-Secretary, 10 May 1935, copy in FO 371/19613 (W 4103/76/50). See also Grimshaw to L.C. Tombs, 18 September 1935, MT9/2501 (M 5008/35).

149. League of Nations A.18.1936.VIII, p. 1.

150. For the draft British text and Grimshaw's report as Chairman of the Committee, see FO 371/19615 (W 9339/76/50).

151. For the draft League convention and replies of governments, see League of Nations C.449.M.235.1935.VIII and A.18.1936.VIII.

152. Article VI (2) of the British draft text stated as follows:

> 'Where an offence is committed within territorial waters, the High Contracting Party in or under the law of whose territory the vessel is registered, shall have concurrent rights of jurisdiction to punish the offence.'

153. Undated French memorandum submitted to the first meeting of the Committee of Experts, FO 371/19614 (W 6728/76/50).

154. Ibid. and FO 371/20478 (W 500/110/98).

155. Mr. Paust joined the second session of the Committee of Experts due to strong representations by the Norwegian Government. Having just denied a request by British shipowners for a place in the Committee, Grimshaw feared the ire of both British environmentalists and shipowners at the presence of a private delegate to the League circle, see his report as chairman, op. cit. note 150.

156. League draft convention, Article II (1) to (4).

157. MT9/2501 (M 14887/34).

158. Ibid., minute of F.W. Daniel, 29 January 1935. The Board was also informed by C.H. Roberts that chemical treatment of oily residues with cresylic acid or other chemicals would make crude oil wastes suitable for road-paving, or at least sink rather than float the oil from ships, see FO 371/19613 (W 4105/76/50). The Admiralty had their own process to use waste oil from ships, see op. cit. note 127.

159. League of Nations A.19.1936.VIII.

160. Moss, op. cit. note 27.

161. MT9/1625 (M 9031/23).

162. *Lloyd's List,* 15 February 1926.

163. MT9/2581 (M 3581/26).

164. FO 371/18498 (W 8175/1324/50. Subsequent reminders were sent in 1927, 1930 and 1931, see MT9/2581 (M 13771/27), (M 14168/30), and (M

10462/31).

165. See MT9/2581 (M 10767) and FO 371/18498 (W 7261/110/98).

166. 46 U.S.C. 391a (1936).

167. Paust to Robert F. Hand, American Steamship Owners Association, 12 August 1936, copy in MT9/2581 (M 10767/36).

168. Letter to the author by Sivert Overaas, Norwegian Shipowners Association, 1 August 1974.

169. MT9/2285 (M 8072/33) and (M 10726/33).

170. British Chamber of Shipping memorandum, 16 July 1934, copy in FO 371/18498 (W 8175/1324/50).

171. Ibid.

172. Joint memorandum of the British Chamber of Shipping and the Liverpool Steamship Owners Association, August 1934, ibid.

173. Chamber of Shipping memo of July 1934, op. cit. note 170.

174. British officials gave considerable weight to an independent survey of pollution round the British coast by Dr. N.K. Adam, a British scientist, in 1934. Dr. Adam concluded that 'The amount visible on the beaches seems far less than might have been expected after some twenty-five years' intensive use of oil fuel, during the earlier part of which few precautions seem to have been taken to prevent discharge of oil.' See *The Pollution of the Sea and Shore by Oil* (London: Published for the Royal Society's private circulation, 25 October 1934), and FO 371/21248 (W 469/469/98) for FO minutes on the study.

175. Grimshaw to L.C. Tombs of the League, 21 April 1936, FO 371/20478 (W 3484/110/98).

176. FO 371/21248 (W 7900/469/98).

177. FO 371/21248 (W 7000/469/98).

178. See FO 371/21248 (W 4571/469/98) and League of Nations A.18. 1936.VIII (10 August 1936).

179. FO 371/21248 (W 8033/469/98).

180. FO 371/18498 (W 8175/1324/50).

181. Cleminson to Grimshaw, 26 March 1936, copy in FO 371/20477 (W 2589/110/98).

182. See *Foreign Relations of the United States* (later cited as FRUS) 1937, Vol. I (GPO: Washington 1954), p. 973. See also German Note Verbale, 15 January 1938, FO 371/22561 (W 958/958/98); information supplied by U.S. Secretary of State Cordell Hull to British FO, 25 January 1938, FO 371/22561 (W 1886/958/98); and French Government Note, 24 February 1938, FO 371/22561 (W 2725/958/98).

183. FO 371/21248 (W 8322/469/98).

184. FRUS 1937, Vol. I, op. cit. note 182, pp. 970-971.

185. See N.A. Guttery's letter to FO, 22 April 1937, citing various excuses they could use, FO 371/21248 (W 7900/469/98).

186. See Lord Stanhope's statement to the House of Lords, 7 April 1937. In 1937, Foreign Minister Anthony Eden, after a Commons debates, had instructed his ministry to expedite matters on oil pollution, see FO 371/19613 (W 5803/76/50).

187. See FO 371/22561 (W 7553/958/98).

188. For the Admiralty's views, see FO 371/22561 (W 9487/958/98); for the Board's views, see FO 371/22561 (W 10761/958/98); and for the FO minutes, see ibid.

189. For the exchanges between British and French officials in August and September 1938, see FO 371/22561 (W 11980/958/98) and (W 16486/958/98).

191. For the League's reply, see FO 371/22561 (W 10761/958/98); for Walters' suggestion, see FO 371/19615 (W 2859/110/98).

192. FO 371/22561 (W 10761/958/98).

193. FO 371/24030 (W351/351/98).

194. See FO 371/24030 (W 3890/351/98) and MT9/5071 (M 6673/39).

195. Ibid.

196. MT9/5071 (M 6673/39).

197. MT9/5071 (M 6673/39).

198. Statement by Admiral H.C. Shepheard before the Subcommittee on Coast Guard, Coast and Geodetic Survey and Navigation, Committee on Merchant Marine and Fisheries, U.S. House of Representatives, H.R. 10886, S.3016, 13 June 1962.

199. Minute of 29 June 1948, MT9/5071 (M 5471/48). Sir Percy Faulkner, K.B.E., C.B., joined the British civil service in 1930, and at the time of the 1954 London conference, he was Deputy Secretary at the Ministry of Transport and head of the Marine Department.

200. *Report of Proceedings International Conference on Oil Pollution of the Sea 27 October 1953* (London: Co-Ordinating Advisory Committee on Oil Pollution of the Sea, 1954), pp. 2-3.

201. UN E/CN.2/68.

202. UN E/CN.2/100 and Addenda 1 to 4 (9 January 1951).

203. ECOSOC Official Records, 13th Session (1951), Supplement No. 4, Transport and Communications Commission, Report of 5th Session, pp. 9-10.

204. See *Ministry of Transport Committee on the Prevention of Pollution of the Sea by Oil, Report to the Rt. Hon. Alan Lennox-Boyd, M.P., Minister of Transport* (London: HMSO 1953).

205. John Crighton, 'Prevention of Oil Pollution and Gas Freeing and Ship Repairing on the Thames,' *Norwegian Shipping News,* 18 November 1959, pp. 1047-1048.

206. Private letter to the author, 10 and 12 May 1973.

207. Private letter to the author, 6 March 1973.

208. Data and quotes are from the Proceedings of the International Conference on Pollution of the Sea by Oil, London, 26 April to 12 May 1954, unless otherwise indicated. When the author did this part of her research in 1973, the records of the conference were still classified and were placed with the British Government and not with the IMCO library. Upon the author's suggestion, British officials of the Marine Division agreed to hand over the proceedings of the 1954 conference to the IMCO library. The author was asked to give the Marine Division library a copy of the report of the British delegation to the 1926 Washington conference, which they did not have at the time, and to have her thesis vetted by British officials in the Marine Division. Further access to British documents yet classified under the British Official Secrets Act until 1983-84 was also granted to the author, provided she read this and made notes only in the presence of a government clerk in the Marine Division office. Secondly, her thesis draft had to be vetted by officials in the Marine Division. It is only now that the author can publish many sections of her original thesis which British officials, who vetted her thesis in 1973, had deemed too sensitive for publication.

209. Author's thesis, op. cit. note 79, p. 224. For the list of participants, see also Final Act of the International Conference, Cmnd. Paper No. 9197 (London: HMSO, 1954).

210. Private letter to the author by a former conference delegate, May 1973.

211. Letter to the author, 12 May 1973.

212. 190 Lords Debates (1 February 1955), pp. 867-870.

213. See UN Conference on Law of the Sea, 1958, UN A/CONF.13/C.2/L. 118, in UN Conference on the Law of the Sea 149, UN A/CONF. 13/40 (1958), summary records of the 29th meeting, pp. 83, 85-87.

214. See author's article, 'The North Sea as a Special Area,' *Marine Policy*, London, January 1978.

215. Counsellor Gunnar M. Boos, K.N.O., R.V.O., served in the Swedish Government in various capacities from 1920-1960. At the time of the 1954 conference, he was Counsellor to the Swedish Board of Trade and chief of the Shipping Department. From 1959-1961, he became an IMCO consultant on shipping and pollution and is known in some circles as the 'eminence grise' of the 1962 oil pollution conference. He has written some ten books on safety of life at sea, trade policy and oil pollution, and was one of the leaders of the Nordiç Council on Oil Pollution, which has about two million members in Sweden, Denmark and Norway.

216. Sir Gilmour Jenkins, K.C.B., K.B.E., M.C., joined the Board of Trade in 1919 and became Permanent Secretary in the Ministry of Transport and Civil Aviation until his retirement in 1966. He served as President of the 1960 Safety of Life at Sea Conference, the 1966 International Loadlines Conference, and the 1954 and 1962 oil pollution conferences.

217. See 86th Congress 2nd Session, Senate, Executive C, 'Message from the President of the United States. Transmitting a Certified Copy of the International Convention for the Prevention of Pollution of the Sea by Oil, 1954, which was signed in London on May 12, 1954, in behalf of certain states but not the U.S.A.,' 15 February 1960 and IMCO Misc (73) 6, 'Status of Multilateral Conventions and Instruments in Respect of which IMCO or its Secretary-General Acts as Depositary,' 31 December 1973, p. 50.

218. IMCO Misc. (73) 6, p. 50.

219. Moss, op. cit. note 27, p. 67.

220. See S.Z. Pritchard, 'Load on Top: From the Sublime to the Absurd,' *Journal of Maritime Law and Commerce*, January 1978; Noel Mostert, *Supership* (New York: Alfred A. Knopf, 1974); and Anthony Sampson, *The Seven Sisters* (New York: Bantam Books, 1975).

221. A. Logan, Deputy Managing Director of Shell Tanker Ltd., 'The Working of the International Convention from the Point of View of British Tanker and Oil Companies,' *Report of the Proceedings International Conference on Oil Pollution of the Sea, 3-4 July 1959 at Copenhagen* (London: Advisory Committee on Oil Pollution of the Sea, no date), pp. 34-35.

222. House of Lords Debates (1 February 1955), cols. 828, 869-870.

223. Remarks of Lord Chesham, House of Lords Debates (1 December 1960), col. 824, and French Note in IMCO OP/CONF/2 (April 1962).

224. See Mouton, op. cit. note 49, pp. 172-173.

225. For the flags-of-convenience issue, see B.A. Boczek, *Flags of Convenience: An International Legal Study* (Cambridge, Mass.: Harvard Univ. Press,

1962); Erling D. Naess, *The Great Panlibhon Controversy: The Fight over Flags of Shipping* (Epping, Essex: Gower Press, 1972); Samuel S. Lawrence, *U.S. Merchant Shipping Policies and Politics* (Washington, D.C.: Brookings Institution, 1966); the report by the OECD Ad Hoc Group on Flags of Convenience, March 14, 1972; and *Maritime Transport 1974* (Paris: OECD 1975).

226. IMCO OP/CONF/2 (29 December 1961), 'Views of Contracting Governments on the Working of the International Convention . . .', pp. 6-11.

227. For the British special measures, see D.C. Haselgrove, 'The Working of the Oil in Navigable Waters Act, 1955,' in ACOPS 1959 Copenhagen conference, op.cit. note 221, pp. 13-22. For the Irish measures, see *Journal of Commerce*, 24 May 1959; and for the Canadian measures, see *Journal of Commerce* and *Lloyd's List*, 18 February 1961.

228. UN ST/ECA/41 (1956), 'Pollution of the Sea by Oil. Results of an inquiry made by the United Nations Secretariat.'

229. 1959 ACOPS conference, op.cit. note 221, p. 61.

230. See ACOPS pamphlet, 'Oil can spoil the beaches of the world,' (1961); *The Glasgow Herald*, 15 December 1961; *Lloyd's List* and the *Journal of Commerce*, 8 March 1961. For Callaghan's statement, see *The Times* and *Daily Telegraph*, 15 December 1961.

231. Unless otherwise indicated, the analysis of the 1962 conference is based on the author's Ph.D. thesis, op.cit. note 79, pp. 273-312, and IMCO records of the 1962 conference (OP/CONF documents).

232. For the French Institut de Peches experiment from 1951 to 1954, see OP/CONF/14 (20 March 1962). For the West German investigations, see 'Report of the Oceanography Committee of the Federal Republic of Germany,' OP/CONF/12 (27 March 1962).

233. For the U.S. statement, see 'Information on Biological Research,' containing Dr. Claude E. Zobell's confirmation of the bacterial degradation of oil at sea, OP/CONF/C.4/Add. 3 (2 April 1962). The history of Dr. Zobell's work and its attractions for oil pollution policy-makers are further described by Moss, op.cit. note 27, pp. 94-98.

234. Letter to the author, 7 July 1974.

235. See author's article on the North Sea, op.cit. note 214, and Part Seven of this book.

236. See minutes of the study group in OP/CONF/C.3./WP.4 (2 April) and 'Circular Letter dated January 1962 sent to ship masters by a U.K. Tanker Company to show the feasibility of the retention of oily residue on board vessels.'

237. Moss, op.cit. note 27, p. 80.

238. These charts were prepared by the West German Hydrographic Institute of Hamburg.

239. Remarks of Dr. Meehean (USA), 2nd Meeting of Committee on Ships, OP/CONF/C.3/SR.2 (28 March 1962). See also summary of discussions of the working group's first meeting, OP/CONF/C.3/WP.4 (2 April 1962), Annex I, p. 2, 'Memorandum of USA Members of the Committee of Ships on Proposal, Article IIA – Discharges from Ships,' 31 March 1962.

240. For the Soviet practice, see remarks of Captain Mikoulinski, OP/CONF/C.3/SR.2, p. 3; USSR Ministry of Merchant Marine, *Manual on the Avoidance of Pollution of the Sea by Oil* (Moscow 1970), also issued as IMCO document OP/IX/3/4 (10 May 1971); M. Nestorova, 'Cleaning of oil tankers with the mulsifying method,' *Ambio Special Report* No. 1, 1972, pp. 123-126; and

M.P. Holdsworth's testimony, *Report of Select Committee on Science and Technology, Minutes of Evidence, Appendices and Index* (London: HMSO, 26 July 1968), pp. 139-140.

241. See 'Revision of the International Convention on Oil Pollution,' *International Conference on Oil Pollution of the Sea, 7-9 October 1968 at Rome Report of Proceedings* (Winchester: Warren and Sons Ltd. [for ACOPS], no date), pp. 280-281; and 677 Commons Debates (17 May 1963) and 251 Lords Debates (28 June 1963).

242. See Proceedings of Rome conference, op.cit. note 241, J.H. Kirby, 'The Clean Seas Code: A Practical Cure of Operational Pollution,' pp. 204-206. See also author's thesis, op.cit. note 79, pp. 337-360.

243. Kirby, op.cit. note 242, p. 211.

244. IMCO Assembly Resolution A.175 (VI), Supplement 1 to 1962 amended Convention, 21 October 1969. The 1969 IMCO amendments took effect on January 1978.

245. A good record of this controversy may be found in the opinions submitted by Counsellor G.M. Boos and M.P. Holdsworth to the *Marine Pollution Bulletin*, November 1970. See also author's thesis, op.cit. note 79, pp. 355-360; and author's article on LOT, op.cit. note 220.

246. For the U.S. delegation views, see IMCO OP/CONF/C.3/WP.4 – Annex I, p. 4, wherein they pointed out that new large ships were being asked to retain 'any oil (not 100 ppm)' onboard.

247. *The Financial Times*, 22 October 1969.

248. For a select sample of this type of literature, see Rachel Carson, *Silent Spring* (New York: Alfred Knopf Co., 1962); Harold Halfrich Jr. (ed.), *The Environmental Crisis* (New Haven: Yale Univ. Press, 1970); Max Nicholson, *The Environmental Revolution* (London, Penguin Books, 1970); Lynton Caldwell, *In Defense of Earth* (Bloomington: Indiana Univ. Press, 1972); the Club of Rome's *The Limits to Growth* (London: Pan Books, 1974); and John Maddox, *The Doomsday Syndrome* (London: Professional Library, 1972).

249. House of Commons Select Committee on Science and Technology, *Coastal Pollution: Report, Minutes of Evidence, appendices and Index* (London: HMSO, 26 July 1968), p. xxv.

250. Richard Petrow, *The Black Tide: In the Wake of Torrey Canyon* (London: Hodder and Stoughton, 1968) pp. 76-77.

251. See Petrow, op.cit. note 350, p. 230; Edward Cowan, *Oil and Water: The Torrey Canyon Disaster* (London: William Kimber, 1969(pp. 202 and 206; *Fairplay*, 6 July 1967; and Norman A. Wulf, 'Contiguous Zones for Pollution Control,' *Journal of Maritime Law and Commerce* 3:3 (April 1972) p. 540.

252. The Labour government issued a White Paper soon after the incident, see *The Torrey Canyon* (London HMSO Cmnd. 3246, April 1967). However, an all-party committee of members of parliament, after conducting a fifteen-month long investigation, issued a highly critical report (Select Committee Report, op.cit. note 249). In a White Paper issued in January 1969, the government rejected the criticism that they had been caught unaware by and had acted poorly in the disaster, see Home Office: *Coastal Pollution: Observations on the Report of the Select Committee on Science and Technology* (London: HMSO Cmnd. 3880, January 1969).

253. For the Liberian report, see Republic of Liberia, *Report of the Board of Investigation in the Matter of the Stranding of the s.s. Torrey Canyon* on

March 18, 1967 (Monrovia, Liberia, 2 May 1967), also in IMCO Council C/ES.III/3/Add. 6 (3 May 1967). For Captain Rugiati's defence, see Cowan, op.cit. note 251, pp. 180-199, and Petrow, op.cit. note 250, pp. 157-158 and 243-250.

254. A.D. Couper, *The Geography of Sea Transport* (London: Hutchinson University Library, 1972) p. 67 and Alcan Shipping Services, *Pollution and the Maritime Industry* (Montreal, Canada, 1972) Vol. IV, p. 43.

255. House of Commons Report, op.cit. note 522, pp. 118 and xx.

256. IMCO Res. A. 246.VII.

257. 1973 International Convention for the Prevention of Pollution from Ships, 1973, Annex I, Regulations 22-25. Later cited as 1973 MARPOL convention.

258. Protocol of 1978 Relating to the International Convention for the Prevention of Pollution from Ships, 1973, Annex I, Regulation 13E (IMCO Sales 78.09.E). Later cited as 1978 MARPOL protocol.

259. IMCO Res.A.146.ES.IV.

260. Protocol of 1978 Relating to the International Convention for the Safety of Life at Sea, 1974 (IMCO Sales 78.09.E).

261. R. Michael M'Gonigle and Mark W. Zacher, *Pollution, Politics and International Law: Tankers at Sea* (Berkeley: University of California Press, 1979) pp. 159 and 182-183.

262. For a good survey, see J. Wardley-Smith (ed.), *The Control of Oil Pollution* (London: Graham & Trotman, 1976) and Annual Reports of the Advisory Committee on Pollution of the Sea, London.

263. *Manual on Oil Pollution* (IMCO Sales 1972.12.E).

264. See Aline De Bievre, 'Shelter from a storm,' *Marine Policy*, April 1983, pp. 125-128.

265. For reports on regional arrangements, see ACOPS annual reports.

266. 1973 MARPOL, Annex I, Regulations 1 and 10.

267. 1973 MARPOL Convention, Article 2(4).

268. Ibid., Article 4.

269. For the Canadian proposal on port state jurisdiction, see the IMCO records of the 1973 Conference, MP/CONF/C.1/WP. 25.

270. The final vote on the Canadian proposal was 16 states in favour, 25 states against, with 10 abstentions.

271. 1954 Convention, Article 11.

272. United States ratification of the 1954 treaty was given precisely with the understanding that 'offences within U.S. territorial waters will continue to be punishable under United States laws regardless of the ship's registry,' see IMCO Misc (73) 6, p. 42. See also opinion expressed in the Alcan survey, op.cit. note 254, p. 461.

273. See MP/CONF/C.1/WP.43.

274. See negotiations and states' votes in M'Gonigle and Zacher, op.cit. note no. 270, p. 209-218.

275. IMO Res. A. 447 (XI).

276. ACOPS Yearbook, London, 1984, pp. 25-26.

277. United Nations Law of the Sea Convention, Article 211 (4).

278. UNCLOS Article 211 (5-6).

279. UNCLOS Article 234.

280. See Protocol of 1984 to amend the International Convention on Civil Liability for Oil Pollution Damage, 1969, Article 3; also Part Six of this book.

281. See R. Maybourn, 'Operational Pollution from tankers and other vessels,' in J. Wardley Smith (ed.), *The Prevention of Oil Pollution* (London: Graham and Trotman Ltd., 1979(.

282. ACOPS Yearbook, London, 1984, p. 46.

283. See U.S. Coast Guard, *Final Environmental Impact Assessment* (Washington, D.C., August 15, 1975), p. 177; J.H. Kirby, op.cit. note 242, p. 209; and M.P. Holdsworth, 'Convention on Oil Pollution Amended,' *Marine Pollution Bulletin,* November 1970, p. 168.

284. M. Kluss, 'Microbiological Action on Oil in the Sea,' *Tanker and Bulk Carrier,* August 1968.

285. See for example Delft Shipbuilding Laboratory's Report for the 1973 Conference, IMCO MP/CONF/INF.15/1. The 1985 National Academy of Sciences Report, which the industry cites as having proven that 'there has been no apparent irrevocable damage to marine resources on a broad oceanic scale by either chronic inputs of oil or occasional major oil spills,' was based on an analysis that noted a 50% decline in pollution from 1975-1985, see *Oil in the Sea: Inputs, Fates, and Effects* (Washington, D.C.: NAS, 1985) and also Exxon Corporation, *Fate and Effects of Oil in the Sea* (New York, 1985).

286. 1973 MARPOL Annex II, Reg. 5.

287. See IMCO MP/CONF/INF. 15/2 (1973); R.K. Roberts, 'The Impact of the 1973 Convention on the Design and Construction of Ships for the Carriage of Chemicals in Bulk,' IMCO SYMP III/5; and G. Stubberud, 'Procedures and Arrangements for the Discharge of Noxious Liquid Substances,' IMCO SYMP III/2.

288. Author's article, op.cit. note 214; see also *North Sea Monitor,* (Amsterdam, August 1984).

289. After the reluctance of Americans to proceed with the 1926 Washington draft treaty, the British government considered a proposal for a modified zonal convention for the English Channel and the North Sea among European states, ironically enough upon the initiative of Harry M. Cleminson, General Manager of the British Chamber of Shipping, see note 107 and minutes of Board of Trade meeting on 17 December 1929, in MT9/2008 (M 15893/29); also in author's thesis, op.cit. note 79, pp. 129-130. For the 1954 Conference proceedings, see author's thesis, ibid., pp. 238-239.

290. M.P. Holdsworth, 'Loading Port Inspection of Cargo Residue Retention by Tankers in Ballast,' IMCO SYMP X/1; and author's LOT article, op.cit. note 220, pp. 214-215.

291. Kirby, op.cit. note 242.

292. ACOPS Yearbook 1984, p. 46.

293. F. Magi and A. d'Addio, 'Consideration of the dimensions of ballast water reception and treatment plants,' IMCO SYMP VII/6.

294. For the IMCO Working Party Report, see IMCO BCH II/WP.5, January 1977.

295. G. Stubberud, op.cit. note 287.

296. Ibid., and D. Tromp, 'Aspects of the concentration of discharged noxious liquid substances in the wake of the ship,' IMCO SYMP III/3.

297. J.A. Gulland, 'World fisheries and fish stocks,' *Marine Policy,* Vol. I:3, 1977, pp. 180-181.

298. J.M. Colebrook, 'Changes in the distribution and abundance of zooplankton in the North Sea, 1948-1969,' *Symposium of the Zoological Society of London*, No. 29 (London: Academic Press, 1972).

299. S.R. Carlberg, 'A five-year study of the occurrence of non-polar hydrocarbons (oil) in Baltic waters, 1970-1975,' *Rapports et Proces-Verbaux des Reunions du Conseil International pour l'exploration de la Mer*, No. 171, 1977, pp.67-69; see also *Soviet News*, 6 April 1976, p.142.

300. Author's interview in London, May 1984.

301. 'Oil, the Dwindling Treasure,' *National Geographic*, June 1974.

INDEX